Uncentering the
Earth

FORTHCOMING TITLES

David Quammen on Darwin and evolution

Daniel Mendelsohn on Archimedes and the science of
the Ancient Greeks

Richard Reeves on Rutherford and the atom

General Editors: Edwin Barber and Jesse Cohen

GREAT DISCOVERIES

WILLIAM T. VOLLMANN

Uncentering the Earth

Copernicus and *The Revolutions of the Heavenly Spheres*

ATLAS BOOKS

W. W. NORTON & COMPANY

NEW YORK · LONDON

Excerpts from *On the Revolutions of the Heavenly Spheres* by Copernicus (translation by Charles Glenn Wallis); *The Almagest* by Ptolemy (translation by R. Catesby Taliaferro); and *Epitome of Copernican Astronomy* and *The Harmonies of the World* by Kepler (translation by Charles Glenn Wallis) reprinted with permission from *Great Books of the Western World*, © 1952, 1990 by Encyclopaedia Britannica, Inc.

For information about permission to reproduce selections from this book, write to Permissions, W. W. Norton & Company, Inc., 500 Fifth Avenue, New York, NY 10110

Manufacturing by RR Donnelley, Bloomsburg Division
Book design by Chris Welch
Production manager: Julia Druskin

Library of Congress Cataloging-in-Publication Data

Vollmann, William T.
Uncentering the Earth : Copernicus and the revolutions of the heavenly spheres /
William T. Vollmann. — 1st ed.
p. cm. — (Great discoveries)
Includes bibliographical references.
ISBN 0-393-05969-3
1. Copernicus, Nicolaus, 1473–1543. De revolutionibus orbium caelestium.
2. Copernicus, Nicolaus, 1473–1543—Influence. 3. Solar system—Early works
to 1800. 4. Astronomy—Early works to 1800. I. Title. II. Series.
QB41.C763V66 2006
520.92—dc22
[B]

2005025864

Atlas Books, LLC, 10 E. 53rd Street, New York, N.Y. 10022

W. W. Norton & Company, Inc., 500 Fifth Avenue, New York, N.Y. 10110
www.wwnorton.com

W. W. Norton & Company Ltd., Castle House, 75/76 Wells Street, London W1T 3QT

1 2 3 4 5 6 7 8 9 0

If we made a model of our galaxy big enough to cover North America, the Earth . . . would be about the size of a large molecule.

—WILLIAM T. HARTMANN (1982)

This book is a member of a series of books about science written mostly by non-scientists. In my own case, the result was an autodidact's exercise in explicating a subject slightly beyond my intellectual competence. Fortunately, an astronomer saved me from myself. Manuscript page 86: "This is not true, unfortunately." Manuscript page 110: "I found this argument hard to follow." (Translation: "This is not true, unfortunately.") If my work is now somewhat truer and more fortunate than before, I have Dr. Eric Jensen to thank.

Contents

Why the Universe Screams

Exegesis: Osiander's Preface and I.1–4

Once upon a time, beneath an unspotted sun • Provenance of the preface • *Rev.* I.1: What ought to be must be • Spherical finitude • I.2: The spherical Earth • Starry proofs • I.3: Proportioning water on the Earth • I.4: Eternal circles, circles around circles • The ecliptic and the Zodiac • The equinoxes • Ecliptic wriggles • A complaint against contrary movements • I.4 (cont'd): "We must however confess that the movements are circular" • On guard.

What We Believed: Cosmology

Centeredness as inevitability • Twelve impieties • Ptolemy's justifications • Polish courtyards • The dead hand • Epicycles

Burnings 221

The Medicean planets • Resolutely Copernican • "How great would have been thy joy" • "Newly emerging values still seeking intellectual justification" • "Safely back on a solid Earth."

List of Diagrams

Notes

In this book, "Earth" refers to the planet we live on, while "earth" is one of the four classical elements. In the interests of consistency I have altered quotations to reflect this usage, just as I have Americanized British spellings such as "centre" (the latter are mostly from translations of non-English works anyhow).

After some hesitation, I decided to employ title case for the Sphere of the Fixed Stars, the Sphere of the Sun, etcetera, to emphasize that most of our protagonists believed these to be real places, or, at the very least, as defined zones at specific distances from Earth. These names do not bear capitals in my primary sources—which are in their turn, however, translations of Greek and Latin originals.

"World" was often used by Copernicus, Ptolemy, Augustine and their fellows to describe the cosmos in which the Earth is set. (Augustine: "By the name of Heaven and Earth" is signified "universally and compendiously, all this visible world.") On the rare occasions when I want to use "world" as a stand-in for "our planet" or "Earth," I do so as you or I ordinarily would, without quotation marks. "World" in the Copernican-Ptolemaic sense always appears within quotation marks, and usually accompanied by a reminder as to its archaic meaning. Where possible, both senses of "world" have been avoided.

Why the Universe Screams

Sixteenth-century science is now as tarnished as sixteenth-century silver. Well, all time-darkened concepts were treasured once; let there be a place for them in the Museum of Ideas! It may even be that if we forgive their grimy irrelevance, sorting and polishing in good faith, a handful might grace our own hoard. But what if they were grimy from the start? The first edition of *On the Revolutions of the Heavenly Spheres* never sold out; and a Vatican astronomer is happy to inform us that a mere fifty years after it appeared, "his tables were superseded by the improved observations of Tycho Brahe . . . At that point, the Copernican work was considered by most astronomers to be obsolete."

"HE WAS A SCHOLAR OF POLISH BIRTH
WHO STOPPED THE SUN AND MOVED THE EARTH."

But he didn't, not exactly; nor was he original in his heliocentric assertion; nor, as that Vatican astronomer so helpfully reminded us, were his celestial delineations perfect. After him, science went on, as it tends to do. Kepler distorted the rotational templates of our already uncentered Copernican cosmos into shapes still less pleasing; Newton swept away fussy half-real geometries, replacing them with *rules* of immense power. And Copernicanism got dusty. Call *Revolutions* a testimonial to the mortality of human effort. With his diagrams and angles, Copernicus will tell you "how great the orbital circles of Saturn, Jupiter and Mars are." But Saturn, Jupiter and Mars refuse to orbit in circles.

So why browse through *Revolutions*'s mathematics-cobwebbed pages?

Newton's First Law, which came into the world long after Copernicus himself was dust, runs thus: **Inertia of motion or inertia of rest will prevail, unless force alters the situation.** Copernicus's lonely labor altered the world's inertia, an inertia of rest by our thinking, of motion according to the thinking of the time—for inertia is always blind. We live no more wisely today than a sixteenth-century European; no more cautiously in our righteousness, no abler to prove for ourselves the architecture of the universe whose various faces science claims to mirror. We know that Mars goes round the sun because we have seen pictorial representations of it doing just that. In Copernicus's time, we knew that Mars went round the Earth because Ptolemy and Aristotle had proven that it did. —Aristotle was the Copernicus of his own age. He thought his way forward, altering the inertia of ancient times. Why was the Earth's shadow upon the moon always circular in every lunar eclipse? Addressing this question, Aristotle became one of the first proponents of a spherical Earth. Observing keenly, reasoning out a sophisticated theory of physical mechanics, ambitiously combining astronomy, mathematics, philosophy and religion, he achieved more than our age holds it possible for any one person to do. Then he died, and his embalmed mummy kept presiding over the banquet.

The Jungians say that "collective symbols of the Self . . . wear out. Religions, convictions, truths, they all age . . . But if the highest values wear out, if they lose their shattering numinous quality, then there is naturally the greatest danger." *Revolutions* was profoundly dangerous in its epoch, and hence profoundly necessary.

When Copernicus finally published, in 1543 (they say that a copy was placed in his hands as he lay dying), the European cosmos had long since grown rotten, its inertia of coherence maintained by two forces: classical astronomy and the Scriptures. Acquaintance with them furthers appreciation of Copernicus's achievement, which is why they will each get their own chapters in this book. The life of the man himself seems of merely decorative interest in a study of how *Revolutions* caused a revolution, but I will try to give him his human due, acknowledging what little we know of his qualities and the circumstantial limitations which he overcame (or not).

As the title of this book affirms, *Revolutions* nudged the Earth away from the center of the universe; it moved the planetary orbits of the solar system into a more accurate configuration; and it demonstrated the immensity of the force required to accomplish both tasks, a force whose units were angles, inferences and solitary years, additiosubtractions, demonstrations and false starts, forty-eight epicycles for the paths of the five known planets, and all forty-eight were wrong, because, moon-orbits arguably excepted, epicycles don't exist!

Copernican force is constructed of sinces and therefores, inclinations and obliquations, ever so many of them now superseded. Just as the concentric circles of a certain diagram (the central circle of which reads "Sol") extend beyond the lefthand margin of his wonderfully neat manuscript, which seems lettered by a professional scribe, so the success of Copernicus projects beyond his careful drudgery, his mistakes, his knowledge and ours. He asserts the axial rotation of our planet, believe him or not; his deductions project beyond their own proofs; it won't be until 1851, 308 years after his publication and death, that Foucault's pendulum will validate him experimentally. He insists that the Earth travels round the sun; F. W. Bessel will prove that in 1838—but first inertia receives its due: Giordano Bruno gets burned alive in the name of the compassionate Christ; Galileo recants our orbiting Earth and lives out an imprisoned lie. The pre-Copernican cosmos has begun to die, you see, so it fights for its life, crushing attackers beneath its whirling spheres. As for Copernicus, he reasons on: "Our problem is to find arc *FC* of half the retrogradation, or *ABF*, so as to know at what distance from its farthest position from *A* the planet becomes stationary . . ." This is how he uncenters the Earth and begins to spin it, with much patience and many errors. He spins it like a rifle-bullet and launches it into the darkness, his target the sacred Unknown. And he finds the arc *FC*, dealing the ancient universe of perfection another dispassionate wound, which is why the universe screams.

Exegesis

Osiander's Preface and I.1–4

Copernicus begins his great treatise with the following traditionalist justification of astronomy: "*What could be more beautiful than the heavens which contain all beautiful things?*" This beauty is Platonic: It will raise up our minds from vice to virtue, and by dint of dazzling us with its complexities, it will make us reflect on God. If this notion strikes you as rhetorical drollery, you probably worship our era's foremost idol, which Marx calls "the cash nexus." Copernicus's era worshipped differently. Indeed, as late as Victorian times we find the astronomer Sir John Herschel advising us that loosening our hold on perceptual and logical misapprehensions "is the first movement of approach towards that state of mental purity which alone can fit us for a full and steady perception of moral beauty as well as physical adaptation." And what for Herschel must have been an Anglican abstraction was in Copernicus's time embraced with Catholicism's literalist zeal. As far as anyone in those days could tell, the heavens really *were* purer, more eternal, less corrupt than the **sublunary realm,** the zone of transience and rot. Heaven was a place above us, and God ruled there. Under these circumstances, how could astronomy not be a form of spiritual meditation?

Once upon a time, beneath an unspotted sun

In spite of the preface's prudent praise of Plato—he seeks to justify his unnerving innovation by exaggerating its tradition-ality—Copernicus stands between Plato and us. "For the motion of a sphere is to turn in a circle," he writes, "*by this very act expressing its form.*" For us this bears scant sense; so long ago did science, philosophy and religion fall out that ref-erences to their prior harmony cannot even evoke sadness. But their ancient alliance, which Copernicanism did its mite to break up, was still operative far into Newton's time. One scholar's survey of the development of physical mechanics concludes: "It would be useless to deny the existence, in the creators of dynamics, of a metaphysics, previous, or simply related, to this science itself. The classical authors of the 17th century were not to depart from this necessity, or if you pre-fer, this servitude." And in Copernicus's century, the sixteenth, that same necessity remained as fundamental as the cosmo-logically centered Earth itself.

In fine, Plato believed in fixed ideal Forms; we believe at least as much in the arbitrary, the lopsided, the random. Why is one of Mars's moons more heavily grooved and cratered than the other, and why are they both triaxial ellip-soids instead of heavenly cue balls? We believe that there *is* no why! Increasingly detailed images of those moons, Phobos and Deimos, have become available to us over time, thanks first and foremost to an invention which came into the world decades after Copernicus left it: the telescope (1610). All at once, it became possible to make out lunar craters, sunspots and other implications of celestial imperfection. Phobos and Deimos continued to orbit Mars undetected until 1877, which is to say a third of a millennium after *Revolutions* was printed. As we consider this book's accomplishment, it would behoove us to remind ourselves, and frequently, that in its era, stars and planets were mere points of light, whose color and magnitude might change, whose place in the heavenly roof could be predicted, not especially accurately—Copernicus improved that situation a trifle—but which remained unearthly in a more than literal sense: almost indiscernible save as presences or absences, hence almost unknowable. What were they made of? Something celestial, no doubt, and probably divine; look how they glowed! Millennia ago, they overshone the civilizations of Greece, Rome, Egypt, Babylon; common sense therefore decreed their changelessness, which further separated them from our mortal world of decay, atrocity and vanity. How could there not be a superlunary realm?

Easy now to sneer at sixteenth-century common sense! But from the perspective of local pragmatism which it served, common sense remained quite adequate. Moreover, as we'll repeatedly see, the craving for order reinforced and was reinforced by its own logic. The desire of any prudent soul to foresee impending good and evil, and if possible to increase the former while averting the latter, made astrology a comfort of which for many centuries we would not endure to be bereft.

Provenance of the preface

So once again: *What could be more beautiful than the heavens which contain all beautiful things?* Thus Plato, Copernicus, Herschel.

I began this book by remarking that any number of time-darkened concepts might be polished bright again to serve as treasures for another era. Might the preface's notion that a particular course of abstruse study can rocket us heavenward be one such jewel? Imagine how our lives would alter if we took that to heart! How many individuals do you know who've selected their vocations for the express purpose of apprehending perfection? (I've met some: artists, devout Muslims, teenaged lovers.) Wouldn't perfection be more in evidence on Earth if a greater number of us devoted our lives to it? "For according to Saint Dionysus," echoes one Pope, "the law of divinity is to lead the lowest through the intermediate to the highest things."

But now I begin to wonder whether that particular time-darkened concept *ever* shed a widespread light! How many peasants in Copernicus's day could possibly have seen their toil as anything but toil? (France, 1483: "The poor laborer must needs pay the wages of men who beat him and cast him forth from his own cottage . . ." Germany, 1480: Desperate gleaners are selling themselves into serfdom. Copernicus was born in 1473, in a Poland whose conditions were similar.) What hope had these unfortunates of ascending from sublunary squalor to celestial wisdom?

One profession did do its best to raise the human level: witch-burners. Then as now, they kept faith with beauty and perfection as they saw them, which is an argument for leaving beauty and perfection in their cobwebbed corner.

Unfortunately, witch-burners cannot be omitted from Copernicus's story.

Indeed, it must have been at least partly on their account that the preface to *Revolutions* got contrived by a well-meaning Protestant named Andreas Osiander, who sought to smooth the book's way, not only with justifications of its elevatedness, but also with assurances of its irrelevance: Don't worry; the sun doesn't really go round the Earth; it's merely mathematically convenient to pretend so!

Osiander's preface was an infuriating surprise to Copernicus, and may well have caused his fatal stroke. It's said that the reassuringly Ptolemaic addition *of Heavenly Spheres* to the title of *Revolutions* was also Osiander's doing, although I see no reason to posit that, for celestial spheres receive many a mention throughout the book; in any event, the preface patronizingly contradicts *Revolutions*'s author, who patently did consider his theory to be more than a mathematical convenience. One historian of science concludes that "Osiander's preface has maintained the ambiguity, as a precaution no doubt. This precaution did not prevent the anathema of the theologians later, but it was not however without utility in propagating the doctrine." In other words, Copernicus died in his bed, not at the stake.

Later we will assess to what extent, how consciously, and for which reasons he imperiled himself through publication. For now, let us return to his treatise—and from here on we'll finally hear his own voice, not Osiander's.

Rev. I.1: What ought to be must be

"The world," he says—and context reveals that by this phrase he means not our Earth, but a larger sphere, including the sun, moon and stars, is indeed—a sphere. It is a sphere "because this figure is the most perfect of all," because it contains the greatest volume, because "everything in the world tends to be delimited by this form, as is apparent in the case of drops of water and other liquid bodies, when they become delimited of themselves." And he goes on in this fashion, employing such antique justifications as "perfect" and "suitable," then underlines his defense of a spherical cosmos with the following analogy: "And no one would hesitate to say that this form belongs to the heavenly bodies."

Again, consider Phobos: a worn, irregular, pockmarked bone-fragment spinning in darkness. Consider our own moon, which, according to my *Cambridge Photographic Guide to the Planets*, is "highly asymmetric."

We are empiricists—in science, at least. As for Copernicus's contemporaries, the narrow limits of observation were merciful and forgiving to their theories, so that they could be proud empiricists of the mind. After Copernicus, there were those who refused to peer through Galileo's telescope because whatever it might show them could not possibly be of value. (As you might have begun to suspect, the telescope is another of our protagonists. It waits menacingly behind the future's black curtain. Every now and then, its lens will let light into Copernicus's story, scorching away another trusted cobweb of the old universe.) Copernicus himself refused to allow that dangerous oxidizer, rationality, to utterly corrode away what was dear to him. Regarding empiricism, let it be noted that during the many years of *Revolutions*'s gestation, its author made only about twenty-seven recorded observations of the sky. So much for quantity (or, as we would say now, for reproducible trials). In quality he proved equally lax. His wooden triquetrum (essentially a ruled stick) gave results whose accuracy failed to exceed ten degrees. One of his acolytes, the young Rheticus, inquired why he troubled not to use the steel circles from Nürnberg which could have brought him within four degrees if he were lucky and careful. He replied: "If I could bring my computations to agree with the truth to within ten degrees, I should be as elated as Pythagoras must have been after he discovered his famous principle."

That was Copernicus for you. As the saying goes, he didn't ask for the moon. He had no reason to. He already knew its nature: spherical, craterless, eternally perfect. What ought to be, had to be.

Revolutions is permeated with this spirit, which will only become more alien to us the more we learn of it. In most respects Copernicus was no Copernican at all, but a devout Ptolemyite. Regarding certain annoying facts which go against the circular orbits of heavenly bodies, he remarked that "since the mind shudders" at the possibility they represented, circularity must be upheld, which he convolutedly did.

Isn't this still the way of the world? A scientist is supposed to follow his observations wherever they might lead, altering his judgments accordingly. Frequently he lives up to this code, especially now that originality (or its simulacrum) is venerated. But does he necessarily act thus in matters of faith, politics, or even in his own field, if doing so might cross his peers? As we shall see, Copernicanism crossed all those lines, and suffered accordingly.

On the larger subject of what ought to be, let's remind ourselves that the purpose of conceptualization is to transform reality's perceptual randomness into patterns. A perfect circle excels in beauty, elegance and mathematical simplicity.

Psychologists have found that we tend to remember circle-like figures as closed and complete, whether they are so or not. And once we accept a circle as our pattern, then at the idea of doing away with it, even on the sadly necessary grounds that it fails to model actuality, "the mind shudders."

Thomas Kuhn remarks about Kepler, who built a mystically mathematical edifice upon a foundation of Copernican geometry, that "today this intense faith in number harmonies seems strange, but that is at least partly because today scientists are prepared to find their harmonies more abstruse." Perhaps any stage of science could make this remark about its own past.

In sum, Copernicus begins his treatise with what we would now consider a faulty premise. The universe is perfectly round because it ought to be. (What shape *does* the universe actually describe? Don't ask me. I refer you to J. D. North, 1965: "It is easy to speak of the infinite . . . but . . . difficult to speak of it meaningfully.")

Spherical finitude

In spite of J. D. North, Copernicus's universe is not infinite at all. In this respect it remains the universe of Ptolemy. Because it possesses a definite shape (the shape of a sphere), it exists within a boundary; it comes to an end.

Even now, mariners and orienteers find it convenient to speak of the **celestial sphere** which rotates over our heads every twenty-four hours. A mid-twentieth-century trigonometrician advises: "Such a sphere, of unlimited radius but with its center at the center of the Earth, is useful in solving certain problems in astronomy and navigation." To Copernicus and his predecessors, this sphere, whose radius was not unlimited at all, but simply undetermined as of yet, actually consisted of nested sub-spheres, one for the moon, others for the various planets. An essential difference between the Copernican system and the Ptolemaic is that in the former the Earth also has a sphere and partakes in these concentric revolutions around a point very near the sun, while in the latter the sun exists within a sphere which partakes in the universal turning about the motionlesss Earth. In either case, beyond the planetary spheres lies the Sphere of Fixed Stars, where all constellations dwell. In Ptolemy's universe they turn in unison, making one westward circuit every twenty-four hours while simultaneously swinging eastward at the more leisurely rate of one degree per century. As for Copernicus, "among our principles and hypotheses we had assumed that the Sphere of the Fixed Stars, to which the wanderings of all planets are equally referred, is wholly immobile." In either case, the Sphere of Fixed Stars is the terminus of created things.

Why can't empty space lie beyond? Aristotle insists that since "nature abhors a vacuum," there can never *be* a vacuum; for his epoch lacked the mechanical apparatus required to create one. And since vacuums are impossible, then beyond the Sphere of Fixed Stars lies no void space, but simply *nothing*. Of course it is possible that we may yet discover other heavenly movements which will require us to posit more spheres—for instance, several good Ptolemaists propose to add a slow-turning sphere to account for the movement called **precession**, which we'll mention farther on—but no matter how we organize our universe, one of these spheres will have to be the outermost. For now, let's leave that position to the Sphere of Fixed Stars.

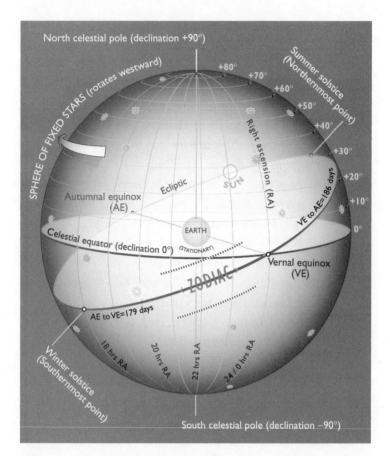

Figure 1 *Where We Are: A Ptolemaic View (Still in Use Today)*
Stars seem to rotate westward about the celestial poles.
Declination = projection of terrestrial latitude, in degrees (°).
Right Ascension = projection of terrestrial longitude, in hours.

Believers in the perfect bygone universe, I ask you this: Is the
Sphere of Fixed Stars an actual entity or simply a theoretical
convenience? In Aristotle's day the various spheres (or shells as
I should say, since in order to move each other and to eliminate

inter-voids they had to be thick enough to contain their planets and the spaces between them) did offer the prospect of their own physical reality; whereas five centuries later, in the time of Ptolemy, they had come to serve more as constructs to keep track of the ever-altering celestial positions. In any event, Ptolemy encourages us to make a model of the Sphere of Fixed Stars, in hue "rather deep so as not to be the atmosphere of day but the night's in which the stars appear," traced with an **ecliptic** circle subdivided into 360 degrees (we'll explain the ecliptic shortly), then marked with constellations' star-points "in yellow or some other distinct color." Whatever the radius of the sphere may be, and successive astronomers will increase it, the fact that it is indeed a sphere requires that every star must lie at the same distance from us. This night-blue globe of heavenly orbs, the farthest thing we can see, is a concretion of harmony, regularity and wishful geometry.

Regarding the absolute size of the universe Copernicus avoids committing himself, but on the subject of its size in relation to that of the Earth he sums up the position of his precursors:

> Many however have believed that they could show by geometrical reasoning that the Earth is in the middle of the world; that it has the proportionality of a point in relation to the immensity of the heavens, occupies the central position, and for this reason is immovable, because, when the universe moves, the center remains unmoved and the things which are closest to the center are moved the most slowly.

In *Revolutions* Copernicus will uphold only one of these assertions, but a very significant one, namely, that the Earth possesses the proportionality of a point.

In the nineteenth century, Sir John Herschel writes of the necessary of "dilating" one's thoughts "to comprehend the grandeur" and vastness of space, after which when one necessarily "shrinks back to his native sphere, he finds it, in comparison, a mere point; so lost—even in the minute system to which it belongs—as to be invisible and unsuspected from some of its principal and remoter members." This humbling alteration in our perception of ourselves has often been blamed on "the Copernican revolution." But obviously we should blame some of it on Ptolemy and company. By Copernicus's time it has long since become received wisdom. Christian theology has absorbed it: Two centuries after Ptolemy's *Almagest*, Saint Augustine writes: "And therefore out of nothing didst Thou create Heaven and Earth; a great thing and a small thing."

What might be the difference between the immensity of Augustine's universe and the immensity of Herschel's? Quite simply, however awe-inspiring the former's dimensions, it becomes in its own turn a mere point in relation to the latter, which is so infinite, or nearly so, that the notion of us marking its center is ludicrous.

At what hour precisely should the foundation-stone of a new church be laid? Ask the astrologer Guido Bonatti and he'll tell you what the stars say—no matter that Dante wrote him into the *Inferno*. Throughout the Middle Ages, many Jews will continue to believe that each of us was born under an allotted star which—always subject to prayer, charity, etcetera—determined our fates. In Herschel's universe, our invisibility and unsuspectedness impel us toward the thought that most stars, the ones we'll never see, have nothing to do with us.

In this regard, Copernicus stands closer to Augustine than to Herschel, as evidenced by the following sentence in *Revolutions*: "I know of no one who doubts that the Heavens of the Fixed Stars is the highest up of all visible things." Or again: "Let *AB* be the greatest circle in the world on the plane of the ecliptic." The prospect that the grandeur goes on in all directions, beyond the greatest circle and the highest up thing that we can conceive, and quite possibly forever, remains literally beyond Copernicanism.

All the same, Copernicus holds up our cosmic insignificance ever before us. Here is a typical reminder from *Revolutions*: "It seems more accurate to say that the equator" of our planet "is inclined obliquely to the ecliptic than that the ecliptic, a greater circle, is inclined to the equator, a smaller."

I.2: The spherical Earth

Not only is "the world" spherical, Copernicus continues, but likewise Earth.

(Ptolemy and Aristotle have already said so. The Muslim scholar Al-Biruni said it again in A.D. 1025, and cited the Qur-'an to prove it. Who said it first? To Anaximander the universe was a sphere and the Earth a cylinder within it, because that way it could play the cosmic axis's part. Anaximenes after him concluded that the sun, moon and stars were the fiery concretions of our Earth's breath; undoubtedly they must be fixed in crystal. Xenophanes obeyed common sense and asserted a good flat Earth. Parmenides and the Pythagoreans decided on a spherical Earth, but they didn't convince the rest of us; for instance,

Democritus insisted that it was a disk. But by the fourth century B.C., Greek thinkers most often asserted, as would Copernicus, a spherical Earth within a spherical "world.")

Nowadays we know that the Earth is actually pear-shaped (so I was taught as a child), or orange-shaped (so Herschel describes it), but this is a trivial gloss—or, as Copernicus would have seen it, a disfigurement. (My high school trigonometry book, published in 1954, advises: "In finding the distance between two points and the direction of one point from another, we assume that the Earth is a sphere with a radius of 3960 miles.") Back to what we all know: Perfection belongs to God. The shapes of the celestial bodies are perfect. The sphere is "the most perfect of all." No matter. From the flat world of common sense to the spherical world of Ptolemy and Copernicus is a huge advance. And our planet is, after all, spherical, within a margin of error of ten degrees! "If I could bring my computations to agree with the truth to within ten degrees, I should be as elated as Pythagoras."

How does Copernicus justify the round Earth? One of his proofs I heard in elementary school: As we watch a ship set out to sea, we find that a pennant on the mast remains visible after the ship itself has vanished over the horizon, then likewise sets like the sun. How could this be, if the Earth didn't curve downward from us in all directions?

Empiricism of this sort remains more plausible than undeniable. For instance (setting various physical and astronomical phenomena aside) one might take the setting of ships to be evidence of a *concave* Earth, in which case the ship's changing angle might increase the apparent protruberance of the mast—a possibility, feeble though it seems to us, which Ptolemy takes pains to disprove.

Other observations support the roundness of Earth. Before Herschel's time, and perhaps even before Ptolemy's, the angular diameter of the visible horizon was being measured by a simple instrument called a dip sector, and the resulting value as seen from a mountain was less than when viewed at sea level. In other words, the higher one went, the more pronouncedly the horizon began to curve down around itself.

But these data led only to logical inferences, not to perceptual certainty. Observation did support *Revolutions* eventually—either 391 or 392 years later, to be precise; for not until 1934 or 1935, from a certain stratospheric balloon which achieved an altitude of 22,066 kilometers, was an image recorded (on infrared film) to show us all that the Earth really does curve into a ball! —All the more credit to Copernicus's predecessors who had reasoned it out.

Starry proofs

Back to *Revolutions*. The Earth is round because it ought to be. Here comes a second observational proof, once again stated more elegantly by Ptolemy: If we travel north, we find that extremely northern stars no longer set, while extremely southern stars no longer rise. In the technical terms which Copernicus reveled in, "the northern vertex of the axis of daily rotation" of those stars' apparent movement "gradually moves overhead, and the other moves downward" and thereby out of sight "to the same extent." Should we travel south, the converse happens. If the Earth were flat, the angle at which we viewed those stars would alter according to our location, but the planet itself could not occlude them. And they obviously do get occluded. This is a very strong argument for a convex Earth at the very least, and very plausibly for a spherical one, since no sea-voyagers have yet discovered any edge; indeed, twenty-one years before *Revolutions* was published, Magellan's death did not prevent us from completing the first circumnavigation.

(On the subject of northwardness, a remark is called for: From our uncentered point of view, the north pole is seen as one of the two endpoints of the Earth's axis of rotation. But by what coincidence did Copernicus's predecessors locate north in the same direction as we? The answer is empirical, and also Babylonian. The shadow of a sundial-stick, or gnomon, will continually alter in accordance with the sun's apparent movement; but at noon each day, which is to say the instant when the shadow is shortest, the shadow's direction will be eternally the same. That direction—in the northern hemisphere, at least—is north.)

To Ptolemy, Copernicus stands likewise indebted to the following: "Add to this the fact that the inhabitants of the east do not perceive the evening eclipses of the sun and moon, nor the inhabitants of the west the morning eclipses . . ." Were the Earth flat, everybody on it could perceive those eclipses, and at the same times.

Furthermore—and like much of *Revolutions*, this goes back through Ptolemy to Aristotle—the terrestrial eclipse is caused by the Earth's shadow on the moon, a shadow which happens to be circular ("perfectly" so, Copernicus naturally goes on to say); therefore, how can the Earth not be spherical?

This section of *Revolutions* adequately represents the whole in its mixture of pieties, deductions and such irrefutable observations as: "So Italy does not see Canopus, which is visible to Egypt." —We might remind ourselves that Copernicus himself never went to Egypt. Why should he? Ptolemy and his forebears took all necessary measurements there.

I.3: Proportioning water on the Earth

Next Copernicus works out how our planet must be divided between water and land. There must be more of the latter, he says, since otherwise we would dwell beneath the ocean. Moreover—and here comes a fine specimen of Copernican obscurity—"spheres are to one another as the cubes of their diameters. If therefore there were seven parts water and one part of land, the diameter of the land could not be greater than the radius of the globe of the waters." What does he mean? The cube root of 7 is about 1.92, half of which hypothetical diameter yields a "radius of the globe of the waters" of 0.96 of a unit. By the same reasoning, halving the cube root of 1 gives us a "radius of the globe of land" of 0.5. Therefore, if the land and the water were both perfect spheres, these values would—yes, we know—put us under the ocean. Frankly, I feel unimpressed with Copernicus's reasoning. Why couldn't the continents rise up tall and steep like sea-urchin spines from a tiny rocky core far beneath the depths? As it happens, Copernicus is correct in his supposition, since although water covers almost 75 percent of the Earth's surface, our oceans go no more than about five miles deep, which works out to a mere 1/792 of the planetary radius.

Here we see one of the striking qualities of *Revolutions*: Often, without sufficient data or even for utterly wrong reasons, Copernicus turns out to be utterly *right*.

He points out that were there more water than land on Earth, the farther out to sea we sailed, the more the bottom would drop off. But one continent and island succeeds the next, so he thinks that the bottom cannot be so very deep. (Again, I offer you my stone sea-urchin.) Indeed, he says (and here his intuition seems especially astute) we keep finding new lands, such as America, discovered fifty-one years before the publication of *Revolutions*. For all Copernicus knows, those new lands could have come to an end in the unknown regions, in which case there might be more water than land after all. But he dares to write: "We would not be greatly surprised if there were antipodes"—for instance, Antarctica, which will not be sighted until 1820.

I.4: Eternal circles, circles around circles

Now that Copernicus has established, at least to his own satisfaction, what the Earth is made of and how it is shaped, he begins to consider the heart of the matter.

"We will recall," he writes, "that the movement of the celestial bodies is circular."

From our centristic vantage point, we see the sun pass by from east to west each day. I myself in my various Arctic summers have enjoyed the spectacle of the never-setting sun wheeling round and round in the sky; if I know what time it was at a given place yesterday, I can note where it is at that same time today, and use it as a compass. (Copernicus: "Where the axis of the Earth is perpendicular to the horizon there are no risings or settings, but all the stars turn in a gyre . . . the horizon coincides with the equator.") Common sense insists that we are still, and the heavens revolve around us. Why shouldn't it be that way, especially since the Bible centralizes our ongoing drama of sin and possible redemption? Parmenides against Democritus is but one mind against another. Heresy against truth is a rather different case!

To the detriment of common sense, the moon goes in the opposite direction from the sun. That's why we good Ptolemyites need to posit that they revolve around us in separate spheres. Or, in the exacter words of my uncentered fellowman, the astronomer Eric Jensen: "The directions are the same. Both show the daily east-west motion, with a more gradual drift to the east against the stars. The *rates* are different (the moon rising about an hour later each night compared to a given star, with the sun only rising about four minutes later), which presumably is why they need their own spheres."

Alas, a few parallel spheres fail to schematize "the world" sufficiently, for the orbits of the five known planets inhabit their own canted plane. Or, in Copernicus's terminology (which here again is Ptolemy's), they follow the **oblique ecliptic.**

The ecliptic and the Zodiac

What is the ecliptic, and why might it be oblique?

Once upon a time, beneath a spotless orbiting sun, Ptolemy presented us with an intuitive and plausible cosmos of two movements: firstly, "that by which everything moves from east to west, always in the same way and at the same speed with revolutions in circles parallel to each other and clearly described about the poles of the regularly revolving sphere." These **celestial poles** are the projections of Earth's own poles, and halfway between them, greatest of the parallel stellar circles of rotation, lies the **celestial equator**, likewise a projection of its terrestrial namesake.

Copernicus has already informed us that depending on our latitude, certain stars will be visible and others invisible; moreover, the closer to each celestial pole a star might be, the less it moves, rises or sets; however, all the stars we do see will rise in the east if they rise at all, and revolve westwards in parallel with this circle, altering their apparent position in the sky by about 15 degrees per hour (since 15 × 24 = one full 360-degree circle), and never altering their position with respect to each other—another good reason for our Earth-centered predecessors to conceive of the starry cosmos as an immense piece of clockwork turning in a perfect circle which gets completed (what a coincidence! I mean, what divinity!) precisely at the end of each cycle of day and night.

In spite of Ptolemy's "everything," there remains an irritating category of celestial objects—the sun, moon and planets—which "make certain complex movements unequal to each other, but all contrary to the general movement"—a phenomenon which Anaximenes seems to have been the first observer to record. Throughout the ages, these orbs have been of greater concern to us than all the others, since, as one history of magic explains, "they exhibit initiative by their individual course, acting according to their own laws and moving in a direction contrary to that of the fixed stars, which travel collectively." Ptolemy accordingly spends much of his *Almagest* on them. They require him "to suppose a second movement different from the general one, a movement about the poles of this oblique circle or ecliptic," which corresponds to the solar line and which literally means "the line on which eclipses may occur." This movement, which really comprises an entire category of complex revolutions in a counter-stellar direction, may be vulgarized as follows: The sun apparently traces a circle around us in the course of a year, while Venus and Mars orbit us in a series of curlicues, crossing and recrossing the sun's path, speeding, slowing, even backtracking (going **retrograde**). The orbit of Mars, while slightly more regular, displays the same sorts of periwigs as Venus's; so do the much more distant paths of Jupiter and Saturn. Thus the planets wander on and off the ecliptic, within a band whose belt of constellations extends about eight or nine degrees to each side.

In the mid-twentieth century, astronomers offered two theories to explain this approximate coincidence of orbital planes: Either Jupiter's immense gravity harmonizes the trajectories of its neighbors; or else all the planets got formed at once. Now the second theory has won out. Either way, their variation from the ecliptic plane is negligible from a Copernican standpoint of accuracy ("if I could bring my computations to agree with the truth to within ten degrees, I should be as elated as Pythagoras"); and *Revolutions* might well have presented them as beads on the same string; fortunately for *Revolutions*'s immortality, Copernicus demands of himself more than ten degrees of rigor in his consideration of this band, which of course is the Zodiac. Astronomers, who in those days were also astrologers, subdivided it into twelve zones of exactly thirty degrees each, rendering it a great clockface of the year in counterpart to the stellar clockface of the night. In tribute to its dozenfold character, Copernicus occasionally refers to it by its ancient name of *dodekatemoria*. Each zone is called after a prominent constellation within it. Thus when we say that "the sun is in Pisces," or "Mars has entered Scorpio" (the Mesopotamians supposed that in the latter case, their king risked being stung to death by a scorpion), we mean that the celestial object in question has wandered into a certain station of the Zodiac. And saying this, we can say more or less what month it is—not precisely, for just as the heavens own wheels around wheels, astronomy boasts complications upon complications, among them precession, the revolution of each celestial pole at the stately rate of one complete circle every twenty-six centuries. (Nowadays we define it as the rotation of the celestial equator as a result of solar and lunar gravitation.)

As it happens, Copernicus's calculation of the precessional period lies within 99.9 percent of our own.

Like the equator, the ecliptic may be considered (ideally, at least) a **great circle**; that is, the intersection of a sphere and a plane which passes through the center of that sphere; hence it comprises one of any number of "equators" which bisects the sphere into two symmetrical hemispheres. Because it is tilted away from the celestial equator at an angle of 23° 27'* (in other words, because Earth's axis leans by just this amount out of perpendicular with the ecliptic plane), Copernicus refers to the ecliptic as oblique.

The equinoxes

When two planes lie out of parallel, they must intersect somewhere, and their intersection will be a line. When the two

---

*I have been advised to explain this notation, which will be employed here and there for the rest of the book. For the subdivisions of a circle into degrees, arc-minutes and arc-seconds, see the section on parallax, below, p. 116. An arc-minute is represented by a single quotemark, an arc-second by a double quotemark. The divergence of the ecliptic from the celestial equator is thus 23 degrees 27 minutes, or about $23\frac{1}{2}$ degrees.

planes each bear a circle around the line of intersection, each circle will intersect the other at two points 180 degrees apart. Accordingly, there must be two points of intersection between the ecliptic and the celestial equator; and these Ptolemy defines as "the **equinoxes** of which the one guarding the northern approach is called spring, and the opposite one autumn." At those two places, which our current calendar marks as 21 March and 23 September, day and night become equal in length all over the Earth. (Keep in mind that at the terrestrial equator, whose projection is the celestial equator, this is so every day and every night.)

And so, what we'd thought to be two points in time become magically transmuted into a pair of points in space. Days and nights become proportions of a circle. This is the genius of pre-telescopic astronomy. It will allow Copernicus to voyage through interstellar thought-space in such a fashion as this: "The duration of days and nights are inversely equal, because on each side of the equinox the durations describe equal arcs of parallels . . ."

Ecliptic wriggles

As simple as the ecliptic may be conceptually (isn't our geo-centric universe made up of perfect circles?), delineating it practically proves formidable, because its apparent position shifts moment by moment, day by day. Copernicus rightly insists that the variable obliquity of the ecliptic should really be thought of as the variable inclination of the equator to the plane of the ecliptic, which never alters. An Earthbound non-astronomer might never know that. As I write this paragraph, looking due south from my home in Sacramento at a quarter

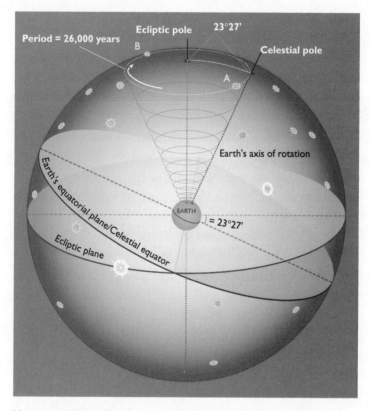

Figure 2 Precession (View is Ptolemaic; Explanation is Copernican)

A and B each become the "North Star" as our axis of rotation passes closest to it. The celestial equator wobbles in perpendicular counterpart to the wobble of the rotational axis, but the angle between the equator and the ecliptic remains fixed.

past three in the afternoon of 10 June 2004, the ecliptic makes an arc high over my left shoulder, curving up and right. This arc flattens and descends dramatically hour by hour; by midnight it will have come about halfway down to my horizon.

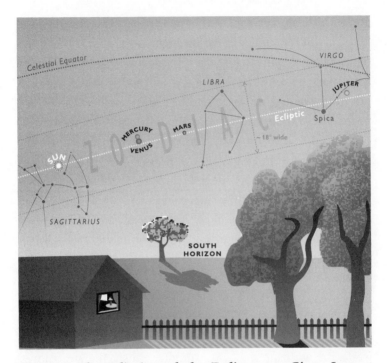

Figure 3 The Ecliptic and the Zodiac at a Given Instant (as seen from Sacramento, California, 28 December 2004, at 10:30 A.M.)

An hour later it will be essentially parallel to the horizon; continuing its clockwise movement, by about six o'clock in the morning of 11 June it will have reached its maximum downward orientation, after which the entire arc, preserving its bend, will begin to rise, tracing out ascendingly concentric semicircles in the brightening sky. By half-past noon it's higher in its southwest corner than it was in the southeast when I began writing this paragraph. (Copernicus: "As the ecliptic is oblique to the axis of the sphere, it makes various angles with the horizon.") At a quarter past three on 11 June 2004, it will return to almost where it was at that time on the tenth, but not quite; for the sun moves eastward through the fixed stars, about a degree per day (fifty-nine arc-minutes in not quite twenty-four hours, to be more exact; and to be still more exact I'd need to note, as Ptolemy and Copernicus will both need to explain away, the fact that the sun's apparent motion is fastest in January and slowest in July), which means that it completes its 360-degree circle in about a year, which length of time is called the **sidereal year.**

Each passing month will be the approximate equivalent of an hour in terms of the ecliptic cycle—but the approximate equivalent only. For example, the summer sun is higher in the sky than the winter sun.

Why must this motion be so complicated? Because the sun and planets partake of both of Ptolemy's two motions. As Thomas S. Kuhn explains it, "Each day the sun moves rapidly westward *with the stars* . . . simultaneously the sun moves slowly eastward along the ecliptic *through the stars.*"

In our uncentered epoch we define the ecliptic in a fashion precisely opposite to the way we did up until Copernicus; it's no longer the great circle which the sun travels around us in a year, but "the great circle in which the plane of the Earth's orbit about the sun intersects the celestial sphere which is to be considered as infinitely great." Knowing this, knowing also that the Earth rotates continuously on its axis, we find some of the ecliptic's perceived peregrinations less in need of complicated explanations than Ptolemy did. Furthermore, we possess, as Copernicus did not, the almost magical advantage of Newtonian mechanics. We are well aware of the concept called gravity, and therefore readily accept our astronomers' dictum that the Earth's orbital plane passes through the line between the center of gravity in the Earth-moon system (a point obviously not the same as the center of the Earth alone) and the center of the sun. This distorts the ecliptic's shape still further; and the further back we can push the limits of observation, the more distortions we can find for any number of subtle reasons.

A complaint against contrary movements

Enough. We've now made the acquaintance of several of Ptolemy's reference points in the night sky; these will also be Copernicus's and ours.

If only the two celestial movements weren't opposed! Then we would have been free to imagine all stars, planets, the moon and even our sun as being fixed on the same revolving moveable sphere. How simple and right existence would then have been at the center of our perfect universe, in the hands of an attentive God.

I.4 (cont'd): "We must however confess that the movements are circular"

Never mind the two movements and the unfortunate distortions of the ecliptic. Set aside the vexities of precession, which alters the equinoctial positions by fifty arc-seconds per year. This is why in the years before *Revolutions*, Copernicus was tortured by "the Eighth Sphere's motion, which the ancient astronomers could not pass on to us in its entirety on account of its extreme slowness." Pretend that the Sphere of the Fixed Stars rotates around us in twenty-four hours exactly; ignore

the maddening fact that each star will rise four minutes earlier tonight than it did the night before.

Even so, our poor "world" turns out to be more complicated than we might have hoped. "For the sun and moon are
caught moving at times more slowly and at times more
quickly," laments Copernicus. The same goes for the five
"wandering stars." "And we perceive the five wandering stars
sometimes even to retrograde, and to come to a stop between
these two movements."

The solution to this worrisome semblance of celestial
imperfection—and it must be no more than that, since "they
maintain these irregularities in accordance with a constant
law and with fixed periodic returns"—is for the heavenly
bodies to execute circles around circles, or, as needed, circles
around circles around circles. "It is agreed that their regular
movements appear to us as irregular, either on account of
their circles having different poles or even because the Earth
is not at the center of the circles in which they revolve."

Circles around circles! It's plausible; it's received truth. It
explains the appearances. Ptolemy says it.

On guard

But Copernicus ends this very revealing section of *Revolutions*
with the following challenge to Ptolemy, to us and to himself:
We must be on guard lest we "attribute to the celestial bodies
what belongs to the Earth." And in spite of his blind loyalty to
circularity, he will be the first to explain the retrograde
motions of the planets, and diverse other whirlings and
twirlings of the heavenly orbs, in terms which will accord
with the supernatural vision of the unborn telescope.

What We Believed

Cosmology

As Copernicus once wrote to the Bishop of Varmia, "Variety usually gives greater pleasure than anything else." Hoping that you share this sentiment, I have chosen to alternate chapters of exegesis with chapters about other subjects.

Accordingly, let us pause again to inspect the evolving universe of Copernicus's predecessors.

Centeredness as inevitability

The development of a near consensus regarding a spherical Earth we told rapidly, because rival views had essentially disappeared centuries before Copernicus wrote *Revolutions*. The *place* of that Earth, and its motion or motionlessness, also partook of consensus—a Ptolemaic one, which Copernicus's mission was to destroy.

Most of the indigenous groups I have studied refer to themselves as *the* people. "The epoch in which Copernicus

lived," not unlike our own, "knew loyalties only toward the
territory one inhabited." In short, we center ourselves within
the worlds we conceive. If we happened to be Venusians, and
could peer through the clouds of sulfuric acid which make
that planet's upper atmosphere so piquant, we would notice
more or less what the inhabitants of Earth always have:

> For they kept seeing the sun and moon and other stars
> always moving from rising to setting in parallel circles,
> beginning to move upward from below as if out of the
> Earth itself, rising little by little to the top, and then coming
> round again and going down in the same way . . . the
> observed circular orbit of those stars which are always visi-

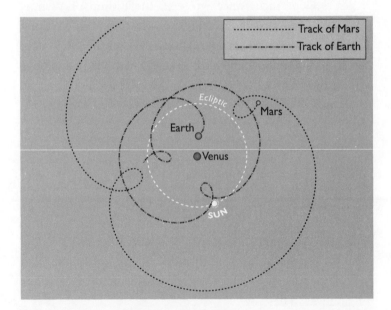

*Figure 4 Apparent Revolutions of Earth, Mars and the Sun as
Seen from above Venus, January 2006*

ble, and their revolution about one and the same center, held them to this spherical motion . . . And then they saw that those [stars] near the always visible stars disappeared for a short time, and those farther away for a longer time proportionately.

(As it happens, on Venus the sun rises in the west, due to that planet's peculiar rotation; moreover, there is no moon; but never mind.)

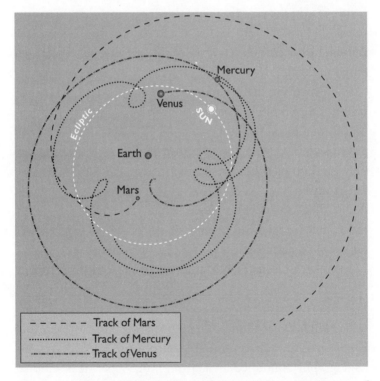

Figure 5 Apparent Revolutions of Venus, Mars, Mercury and the Sun as Seen from above the Earth, November 2005

Wherever we were, why *wouldn't* we place ourselves at the hub of the universe? Perhaps we might choose to define the center as the celestial pole around which all stars seem to turn. But even that is no more than the nearest pole of the observer's planet, projected straight up into the apparently rotating firmament. No matter how we consider the matter, intuition encourages us to center the "world" around ourselves.

Twelve impieties

Hence it's (in the Marxist cliché) "no accident" that three centuries before Christ, hence eighteen before Copernicus, Aristarchus of Samos gets scolded for impiety and error: He's dared to say that we orbit the sun!

Did I mention that his name gets crossed out of the manuscript of *Revolutions*? It may well be that Copernicus had second thoughts about appealing to this particular classical example. So I've read. On the other hand, the deletion occurs only on folio page eleven; in other places, Aristarchus's name remains. What are we to make of that? Was the excision a sly dodge, like Osiander's preface, meant to send away happy those fault-finders who never read much beyond page twelve? Or was it meaningless?

In any case Aristarchus was the first, according to one Copernican scholar. Or, if you prefer, Aristarchus wasn't the first. Shortly before the close of the fifth century B.C., a certain Philolaus or Philolaos preposterously theorized that Earth turns round a central fire which burns beneath us. One hagiography announces that "from him, Copernicus adopted the idea that the Earth moves."

The Ptolemaic consensus, when it did come, was by no

means totalitarian; heliocentrism is said to have been "mentioned by at least twelve philosophers from Plato to Copernicus," and open-minded persons sometimes took note of them. Near the beginning of the twelfth century, for instance, we find a

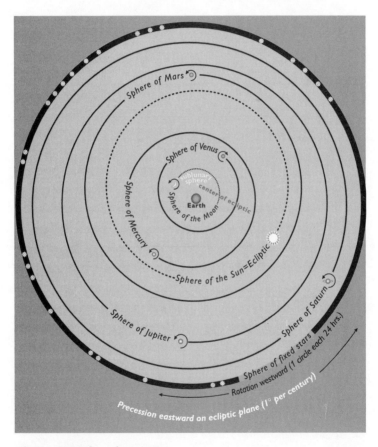

Figure 6 Ptolemy's Cosmos
Scale: Unknown. Order of Spheres of Venus, Mercury and Sun still debatable. Epicycles and equants omitted for simplicity, but direction of epicyclic eccentrics indicated.

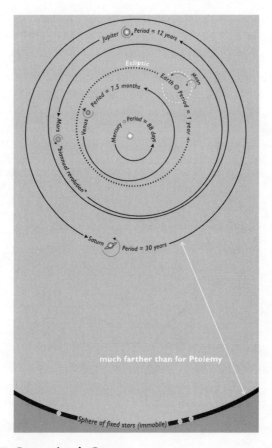

Figure 7 **Copernicus's Cosmos**
Relative planetary distances now approximately calculated, like-wise planetary periods. Copernicus's cosmos resembles ours—if we lacked telescopes and if we hadn't jettisoned the Sphere of Fixed Stars and uniform circular motion.

In this simplified version, eccentrics are indicated, epicyclic motion merely suggested. (Note that the arrows turn in the opposite direction of Ptolemy's.) As usual, I have left the bothersome Mercurian motion alone.

certain Abelard of Bath, just returned from his studies of natural science among the Arabs, expounding on various "Questions on Nature," the fiftieth being "how the Earth moves."

But about Ptolemy's great work one commentator remarks that "its perfection is such that it often covers up the modes of discovery, and its geocentric theory is propounded with only the barest references to its heliocentric opponents."

Ptolemy's justifications

What then were Ptolemy's arguments for centering us?

The Earth *must* be in the middle of all things, or the horizon wouldn't cut the night sky exactly in two as it does.

If the Earth were off the axis around which the universe turns, equinoxes would occur at different times or not at all. Remember that Ptolemy has already defined the equinoxes as points halfway around the apparent ecliptic circle, which certainly adds plausibility to the assertion that if the Earth had been placed off center with respect to that circle, the two intervals between the equinoxes would be unequal. (By the way, who cares when the equinoxes fall? Not least, medieval Jewish astrologers, because at those times, and at the solstices, too, water was said to become poisonous.)

Next defense of geocentrism: If the Earth were, say, east of the universal axis, stars in the east would look larger than stars in the west—a disappointing claim, since Ptolemy has already admitted that the universe is immense enough, the stars far away enough, that he cannot even measure the **parallax** of any star (this term we'll discuss in due time, when we consider the orbit of Venus). Why then should anyone expect to perceive variations in stellar size as a result of *positional* variation? But—and this point must be continually reiterated—neither Ptolemy nor Copernicus will ever comprehend how vast the universe actually is.

Polish courtyards

Each development in the history of astronomy resembles another of the arched courtyards of Copernicus's epoch. We pass through necessity's narrow doorway, enter another open-skied realm of freedom, and yet we're still not really outside; error continues to wall us in—until the very end, when Copernicus, Kepler and Newton wreck the universe; then the walls come down, and we find ourselves alone in Herschel's infinite darkness.

While they last, these walls have a tendency to reify themselves into spheres. Anaximenes is said to have first envisioned the transparent sphere to explain the revolving heavens. Then come the twenty-six spheres of Eudoxus (who posited spheres within spheres), the thirty-three spheres of Callipus, the fifty-two spheres of Aristotle . . . Each courtyard gives way onto another archway, another courtyard.

In the scheme of Parmenides, we dwell within fire; then come the moon, stars and sun together; beyond them, more fire; and at last the outer skin of the universe.

The Pythagoreans are credited with the model of celestial circularity; but who knows on whom to bestow the credit of primacy? In Plato's *Timaeus* we read that the creator of the universe "made seven unequal circles having their intervals in ratios of two and three . . . and bade the orbits proceed in a direction opposite to each other." The sun, Mercury and

Venus were made to rotate at equal speed, while the moon, Mars, Jupiter and Saturn were commanded "to move with unequal swiftness to the other three and to one another, but in due proportion."

How do we know that the Sphere of the Fixed Stars must be farther away than the planetary spheres? Copernicus gives the ancient answer: Because the stars twinkle and the planets do not.

Copernicus will continue to be faithful to the old cosmology in his fashion, positing circular orbits and spheres. Retracing our collective steps, he searches through the past's courtyards for the treasures he prizes above all: celestial observations. Does he sometimes imagine himself back there? He murmurs to us: "At this time Canicula was beginning to rise for the Greeks, and the Olympic games were being held, as Censorius and other trustworthy authors report."

The dead hand

I keep reminding myself that Copernicus lacked a telescope. In place of that inconceivable instrument, he referred to Ptolemy's *Almagest*.

What moves me the most about *Revolutions* is the struggle it represents to free the human mind from a false system—Ptolemy's system. You know by now that Copernicus will liberate himself only partially from that system, which Sir Richard Woolley once called "the longest tyranny." Woolley spoke truly: For fourteen hundred years we've believed in Ptolemy's Earth-centered cosmos of revolving spheres! Half a millennium before Ptolemy, Aristotle proposed much the same idea, and who's puffed up enough to claim that Aristotle isn't always right?

Accordingly, I began this book with an animus against Ptolemy. But as I dipped into *The Almagest*, I began to understand how much Copernicus and all of us owe to that indefatiguable cataloguer of stars, theorist of music and of optics, geographer and geometer. When Ptolemy explains what the ecliptic is, I comprehend his explanation much more readily than those of many modern-day astronomy textbooks. Copernicus for his part rarely troubles to define terms. He assumes that we have all read *The Almagest*.

The single most admirable thing about this masterpiece is that it begins with observations and interprets them according to geometry. Of course such precursors as Hipparchus did the same, but *The Almagest* approaches completeness as well as consistency, exemplifying almost perfectly the sort of quantitative rationality which transforms sensory phenomena into entities subject to mathematical manipulation. At the end of the twentieth century, one astronomer defines the phases of the moon as follows: New moon, "first quarter, full moon, and last quarter refer to the moments when the celestial longitude differences between the moon and sun are, respectively, 0°, 90°, 180° and 270°." This way of thinking, which we've already met in Ptolemy's geometric definitions of the equinoxes, encourages us to visualize intermediate numbers, draw a diagram, and see in precise proportions what we have already seen in intuitive harmonies. It is our legacy from the ancient astronomers, of whom we should single out Ptolemy: His tyranny was the longest because it was the most perfect.

He retains admirers to this day. One historian of ancient astronomy and mathematics, after having insisted that Copernicus's theory bears such a close resemblance to Ibn al-

Shatir's—as indeed is the case—that "independent discovery is quite out of the question," gives the knife one more dismissive twist: "I must emphasize that once one has suggested a helio-centric solar system, one can immediately find its dimensions in **astronomical units*** from the *Almagest*'s parameters."

By current standards, the figures of both Ptolemy and Copernicus are often off the mark. For instance:

Some Relative Celestial Diameters

	Ptolemy's calculation	Copernicus's calculation	Current estimate
Moon	1	1	1
Earth	"Very nearly 3 and 2/5"	1.35	1.84
Sun	18 and 4/5	24.3	400.02

When we remind ourselves that their determinations were made by eyesight, mindsight, brass circles and ruled beams, these errors become less objectionable. All the same, I see good reason why Ptolemy and Copernicus are read mainly by historians, whose task it is not to compute the true solar diameter but to assert with the appropriate level of tendentiousness that somebody's independent discovery of something is or is not out of the question.

Indeed, so far we've found very little independent discovery in *Revolutions*. Another man's preface, a received idea or two, what does that amount to? "The world" must be a sphere, writes Copernicus. A quick cross-check against the

---

*An astronomical unit is a standard of measurement based on the eclip-tic radius. The mean distance of the Earth from the sun is 1 AU, or around 150 million kilometers.

great text of his dead rival proves him, as he wanted to be, reliably derivative.

Ptolemy reasons more spaciously than Copernicus, more clearly, more beautifully if I may say so, and arguably with fewer crabbed complications. He remains a better thinker than Copernicus in every respect except for the most important one of whose thought is most true.

Epicycles

Much of the history of science consists in this: *Observation slowly overcomes intuition.* The stars and planets turn around us, yes; but round about 150 B.C., Hipparchus of Rhodes, whom both Ptolemy and Copernicus will eulogize, proposes a cosmological center *near* Earth rather than precisely on or in it, because in the northern hemisphere, winter lasts 178 days while summer lasts but 187. Such non-geocentric celestial circles are called **eccentrics**. My entirely uncentered sensibility perceives them as sad compromises in a misguided cause; in other words, they deserve accolades for preserving us at the heart of everything for centuries longer. Eccentrics lay in a convenient pocket of Ptolemy's bag of tricks; Copernicus will employ them, too. How does he explain the seasonal peculiarity which Hipparchus noted? By positing the exact inverse, namely that "the Earth in its annual revolution is not absolutely around the center of the sun." Well, why not? Intuition can still have its due; Hipparchus's new whimsy does not stand in the way of the sun's passing under us at night; and Copernicus need not abandon the verity of perfect circularity.

Ah, circularity! Well, it's necessary in the case of celestial bodies, without a doubt, and it reveals itself in all its elegant perfec-

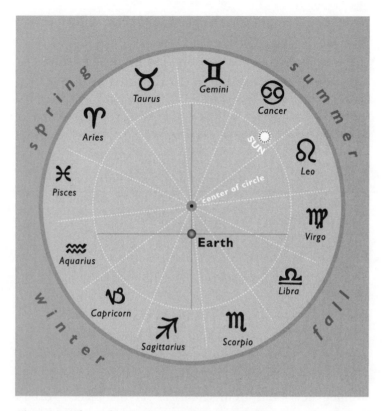

Figure 8 The Ecliptic as an Eccentric Circle
We now know that the Earth travels elliptically (and at varying speed) around the sun. To save the old Earth-centered universe with its credo of uniform circular motion, we can express "the appearances" by placing the sun on a circle centered near to but not on the Earth.

tion in the case of what Ptolemy calls the Prime Movement, the daily rotation of the Sphere of Fixed Stars from east to west, which takes place around the celestial poles.

But what about the Second Movement, or *Movements*, as I

should say? I mean, what about the jitterings of the planets? How can our intuition of circularity be satisfied in their case?

Ptolemy and his predecessors find a way—or, from our uncentered perspective, they make a fateful error. It is natural and seductive; any one of us might have made it. Indeed, the very mathematically minded astrohistorian Asger Aaboe (who tends to praise Ptolemy and denigrate Copernicus) remarks that "a simple epicyclic model is a very reasonable method of accounting, in first approximation, for a planet's behavior when viewed from the Earth." "A simple epicyclic model!" How simple is simple, exactly?

The principle of Occam's Razor—the least complicated hypothesis which can explain all the facts is most likely to be true—has not yet been coined, but in his own fashion, Ptolemy attempts to follow it in *The Almagest*: ". . . it is first necessary to assume in general that the motions of the planets in the direction contrary to the movement of the heavens are all regular and circular by nature, like the movement of the universe in the other direction." In other words, we must not be misled by any seeming deviations from circularity in the planets' orbits, because the universe is rational, logical, elegant. (Doesn't simplicity itself bid us to think so?) Accordingly, Ptolemy's solution to the deviations is rational, logical, elegant: If the orbits of the planets are "considered with respect to a circle in the plane of the ecliptic concentric with the cosmos so that our eye is in the center"—that is, considered from the vantage point of a geocentric universe—"then it is necessary to suppose that they make their regular movements . . . with other circles borne upon them called **epicycles**." If the orbit of Venus appears irregular, then by all means postulate as many circles within circles as geometry requires to make it regular!

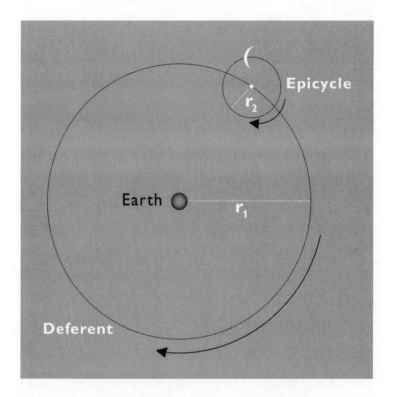

Figure 9 Epicycle (Ptolemaic and Copernican View)
*This simple epicycle depicts the moon. Both revolutions must be
perfect circles. The diagram is for didactic purposes only. Both
Ptolemy and Copernicus added complexities to their descriptions
of the lunar orbit. (Not to scale.)*

So there it is: Venus's invisible gyre within a never to be seen
sphere which revolves divinely and eternally upon the ecliptic
plane, which in its own right turns eastward about one degree
per century (this was *The Almagest*'s calculation of precession,
remember). For Venus, and indeed for the other four planets

as well, Ptolemy has applied the following hypothesis: "The eccentric circle [or deferent, which] the epicycle's center is always borne on is described with its center at the point bisecting the straight line between the centers of the ecliptic and of the circle effecting the epicycle's uniform revolution," the latter being the equant.

Apollonius (220 B.C.) is said to have been the one who conceptualized epicycles. But I fail to see why it couldn't as easily have been somebody else. Before Hipparchus, Eudoxus (d. *ca.* 355 B.C.) had demanded that these circles within circles, or spheres within spheres as I should say, be centered about the Earth, twisting simultaneously in different planes to give the required planetary wriggles. Unfortunately, planets sometimes vary in brightness, which implies that their distances from Earth vary likewise. Ptolemy himself points out that if stars appear to alter their distance from the Earth, their orbits around us cannot be spherical; and this is precisely the case with the "wandering stars." Never mind. Hipparchus, Ptolemy and the rest eventually propose that each planet travels in a circle which is perfectly centered upon the circumference of a larger circle, which in turn is centered upon the Earth. The larger circle is called the **deferent**; the smaller, the epicycle. Ptolemaic astronomers assume firstly that both of these circles lie upon the ecliptic plane, secondly, that they partake of the Prime Movement; and thirdly, that they twirl in the same direction.

Another way of interpreting those same planetary irregularities would be to postulate that their orbits trace circles which are eccentric to the ecliptic. Ptolemy is willing to apply each hypothesis to what appears to be the most appropriate context: **anomalies** of the planets with respect to the parts of the Zodiac are best explained by eccentricity; anomalies of the planets with respect to the sun are better fitted to epicycles. (An anomaly is a regular motion which in combination with another regular movement causes the latter to seem irregular.) Accordingly, Copernicus will comment in *Revolutions* that "the ancients found in all the planets two digressions in latitude answering to their twofold irregularity in" celestial "longitude—one digression taking place by reason of the eccentricity of the orbital circles, and the other in accordance with the epicycles."

Do epicycles exist or not? Well, the orbit of the moon around the Earth is a classic example of an epicycle, and both Copernicus and Ptolemy see it as such. (Please set aside the fact that *Revolutions*'s account of the lunar orbit requires first a retrograde epicycle moving uniformly on a deferent, and then, circling round this, a much smaller direct epicycle with a different period. Who said simplicity was simple?) And if the planets likewise revolved around us, epicyclical theory would certainly explain their movements. Talk about cosmological paradise! Then the universe could easily be comprised of Plato's eight diversely colored celestial spheres, revolving around the spindle of Necessity, with one singing Siren inhabiting each sphere, uttering her individual note now and forever, so that she and her seven sisters create the Music of the Spheres.

It has been said that "in Ptolemy's universe, mathematics merged with moral philosophy, allowing the mathematician to function as philosopher, all the while striving to imitate the divine"—in which case, empirical verification is scarcely the point, so who am I to carp at epicycles? Even Copernicus couldn't get by without them. The twentieth-century astronomer Sir Bernard Lovell enthuses: "This brilliant device, involving only three perfectly regular circular motions, thus easily explains the occasional retrograde motion of the planet."

Indeed, with these circles we can do more than that. Track the path and angular velocity of a celestial orb, construct circles of the size which observation and geometry require, and someday, when Copernicus comes along, it will turn out that you built better than you knew. The radius of Ptolemy's Martian epicycle is $39\frac{1}{2}$ units (or, as he expresses it, 39 theta 30'); the radius of his eccentric circle for the same orb is 60. Now, if we uncenter ourselves in obedience to the compelling circles and angles of *Revolutions*, we'll come to see that the eccentric radius of any planet equals its relative mean distance from the sun, while the epicyclic radius corresponds to Earth's relative mean distance from the same point. Never mind the fact that Ptolemy's eccentric radii for all four planets (and the sun) equal 60 units while the epicyclic radii vary; this is simply an artifact of observations taken from a moving Earth rather than a relatively motionless sun. The important fact is the ratio itself. For Mars, then, the ratio is 60 divided by $39\frac{1}{2}$, or 1.518, a number which differs by less than 1 percent from the currently calculated mean Martian distance from the sun of 1.524 astronomical units.

In short, Lovell was right, and the ancient circles do possess, if not empirical verification, empirical consonance. Why not believe that one more circle, or a hundred more, will perfectly explain the revolutions of the heavenly orbs at last? No wonder that Hipparchus even invented a delicate little epicycle for the sun's path . . .

Diagram of a water-mill

I have before me a mechanical drawing in cross-section of a water-lifting device, *circa* 1615, or just about when *Revolutions* got banned. About a central gear-circle (which ironically resembles the sun), a knurled ring whose lines remind any good Newtonian of motion vectors has been attached by spokes which were visible in the previous perspective drawing, not in this one. Because three dimensions have been necessarily reduced to two for the cross-section, another geared ring which is actually parallel to its knurled cousin seems to be mounted directly on it and evidently turns about its circumference as if traveling about the central sunlike gear.

When I first saw this image, my reaction was: an epicycle! And this epicycle appears to obey the physical laws I believe in. Can't toothed gears cause each other to rotate? If we posit, as Ptolemaists had to, invisible wheels turning one another within the celestial ether, an epicyclic cosmos takes on as concrete a realism as any diagram of a water-mill.

Equants

After further puzzling over the anomalies of the five known planets, Ptolemy takes another step away from simplicity, from Aristotle and ultimately from truth, concluding that "the epicycles' centers are borne on circles equal to the eccentrics effecting the anomaly, *but described about other centers*. And these other centers, in the case of all except Mercury, bisect the straight lines between the centers of the eccentrics effecting the anomaly and the center of the ecliptic." In other words, not only has the center been moved from where intuition places it—on Earth, with us—to whichever point will most plausibly compensate for observational discrepancies, as occurred in the eccentric solar circle of Hipparchus; but now the motion of the ·planet is unvarying not as far as that center is concerned, but with respect to some new point, calculated geometrically, from whose vantage alone the sweep of the planet around us can be made unvarying. Does Mars seem to speed up and slow down in its Earth-centered whirl? We can't have that! So let's posit one center for its orbit, and another for its *uniform speed.* (Here let's note one difference between Copernicus and Ptolemy: *Revolutions* will insist on uniform speed about a physical geometrical center,

Figure 10 Equant (Ptolemaic View Only)
This is a generalized diagram for the perceived geocentric revolutions of all the known planets except Mercury, whose divagations I spare you.

For convenience, Ptolemy assumes all three circles to lie in the ecliptic plane, although considerations of latitude would require us to incline the epicycle in relation to the two eccentric circles, which would in turn be canted off the ecliptic plane.

The radius of the epicycle is r_2. The other two circles are of equal area; that is, their radii are both r_1 units long.

The center of the deferent bisects the line between the equant (center of the apogee circle) and the Earth (center of the ecliptic circle).

The planet's epicycle gets carried round the deferent as usual, but the epicycle's diameter always points to the equant. The planetary motion is uniform only with respect to the equant.

(Scale: Variable and unknown.)

even if that center is but the circumference of a deferent; whereas *The Almagest* demanded uniform speed only in relation to its own mathematical point.) Somewhere, from some perspective, Venus must be passing through the same num-

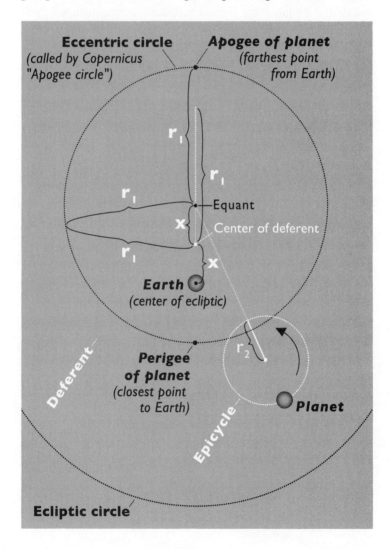

ber of degrees of arc each month. We will determine that "somewhere" geometrically. For we insist that Venus's orbit remain circular.

One way of elucidating epicycles and equants would be to say that the planet concerned *oscillates around a mean position.* Isn't frame of reference almost everything? If epicycles and equants were able to serve (quoting Aaboe again) as reasonable approximations from a pre-telescopic Earthbound frame of reference, why cavil?

R. Catesby Taliaferro, who translated my edition of *The Almagest*, points out that

> this broadening of the principle of celestial mechanics to allow an epicycle's center to move regularly about a point other than the center of its deferent was for Copernicus the major scandal of the Ptolemaic system and one which his own allowed him to remove only at the expense of the appearances.

Accordingly, Copernicus will stand firm. He'll allow no equants in *Revolutions.* "And so there were three centers," he rails against Ptolemy's Mercury, "namely that of the eccentric circle carrying the epicycle, that of the small circle, and that of the circle which the moderns call the equant. They passed over the first two circles and acknowledged that the epicycle did not move regularly except around the circle of the equant, which was the most foreign to the true center, to its ratio, and to both the centers already extant. But they judged that the appearances of this planet could be saved by no other scheme . . ." In other words, "the mind shudders."

He was hardly the only diner who choked on equants. Half a millennium before *Revolutions*, we find the great Muslim scientist Al-Biruni complaining that in some quarters it appears

> possible to imagine a variable speed for the sun about the center of the universe, and a uniform speed about a different center ... and ... possible, also, in the case of the planets, to imagine variable speeds for the centers of their epicycles, on the circumferences of their eccentric deferents, and uniform speeds about the centers of their equants. If all that is possible, then a very serious criticism can be levelled at the moral integrity of those people,

"those people" being the Ptolemaic astronomers.

In my own opinion, Al-Biruni was right; equants were a shabby device. But you can't either please or offend everybody, especially not in the history of science. Jacobsen, for instance, writes Copernicus this epitaph: "His results were for the most part no more accurate than those of Ptolemy, but in one sense they were inferior, for he abandoned the principle of the equant in favor of uniform motions."

The parable of the Alphonsine Tables

Let's speak of accuracy. In the same breath, let's speak of simplicity.

Possessing or lacking integrity the Ptolemaists may be. All the same, they do keep sailors and astrologers happy by predicting the courses of the "wandering stars" with some degree of accuracy. Intuition has almost caught up with observation.

Do you wish to know where the sun will be at any given time? Ptolemy can help you figure that out. "And so we shall arrange the table of the sun's anomaly into forty-five rows and three columns." Then what? Industriously reckoning up ancient observations of equinoxes, he concludes that in the course of those 879 Egyptian years, 66 days and 2 equatorial hours, the sun has traveled a certain mean distance, which he plausibly refines and corrects according to the geometry of equinoctial points upon the ecliptic circle, with its attendant epicycles. Then,

> whenever we wish to know the course of the sun for any desired time, taking the total time from the epoch to the proposed date with reference to the hour in Alexandria and taking it to the tables of mean movement, we add the degrees corresponding to the particular numbers to the 265°15' of the distance found above, and striking the complete circle, we subtract the rest from the 5°30' within the Twins backward [that is, from west to east instead of from east to west] in the order of the signs [of the Zodiac]. And wherever the number falls, there we find the mean course of the sun.

This is where Ptolemy's loyalty to simplicity has led him. But the universe itself was never simple; and any formula counts as a victory if it is practical, which is to say effective within the limits of observation, even if it remains founded on those very errors which Al-Biruni found so odious, such as the notion that the center of the sun's epicycle speeds up and slows down at different points of the ecliptic. As we've noted, apparent solar motion does vary.

In any event, Ptolemy's hope that the planets' asymmetric curlicues of apparent movements can be plausibly translated into perfect real movements seems often just about to realize itself. Against the dreary paragraph just quoted let us set this following equally typical sentence from *The Almagest*: "The angle at *B* embracing the star's regular passage on its epicycle" (remember that for Ptolemy, planets were merely a troublesome subcategory of stars) "is always the difference between the angle at center *F* which embraces the star's regular movement in longitude and the angle at *E* which embraces its apparent movement . . ." In short, the difference between what we see and the Perfect Real is but an angle or two. This simplicity, or near simplicity, or the illusion thereof, renders equants tolerable to "those people."

But as the centuries revolve around our motionless Earth, Ptolemaism becomes ever less simple. By 1252, when King Alphonso X of Castile sets astronomers to work drawing up present and future planetary positions, they require a full decade, and employ so many circles upon, around and within circles that the King awards them this epitaph: If he'd been at God's shoulder during the Creation, he'd have prayed for a simpler configuration.

One thing with many effects

Hence when Copernicus writes *Revolutions*, we're still lost in a time when, as he says at the very beginning of Book I, "the courses of the planets and the revolution of the stars cannot be determined by exact calculations and reduced to perfect knowledge"—when we disagree on the quantity of required deferents and epicycles*; when, indeed, even the length of the year remains unknown. Much of *Revolutions*'s purpose will be to demonstrate the superior simplicity of heliocentrism for determining all those matters—for haven't Ptolemy and the Alphonsine Tables fallen short? The exasperated Copernicus invokes Occam's Razor: "We should rather follow the wisdom of nature, which, as it takes great care not to have produced anything superfluous or useless, often prefers to endow one thing with many effects."

*Here's another telling parable for you: In the thirteenth century, Albertus Magnus makes his astronomical universe out of twenty-six deferent spheres, his astrological cosmos out of only ten.

Exegesis

I.5

"Now that it has been shown that the Earth too has the form of a globe," pursues Copernicus, "I think we must see whether or not a movement follows upon its form." Of course almost everybody agrees that the Earth "rests in the center of the world"; all the same, he continues with his usual understated caution, "the question has not yet been decided and is by no means to be scorned."

In order to appreciate his literally revolutionary argument, and the fixed position of his opponents, our brief narrative of cosmological development is not quite sufficient. We also need to remember that the physics of motion as everyone then understood it was as different from the Newtonian mechanics we now routinely utilize as are pantheistic cults from institutional monotheism; indeed, the history of motion theory is a progression from multiplicity to unity. Some of this history must now be summarized.

What We Believed

Motion

Water flows downhill, toward the ocean. Rocks fall to the earth they came from. But flames strain upward in the direction of the stars; air also rises, as we can see from a swimmer's bubbles. It is therefore intuitively evident that each of these four elements partakes of its peculiar nature, returning to its own place by inclination. Isn't there some essential fire-ness which impels the flame's ascent, some antithetical water-ness which prevents spilled wine from doing the same? Observation and common sense both militate against any so-called theory of "gravity," because flames and air don't fall!

From these self-confirming observations, a longstanding theory of motion came into being—and not only motion, but alchemy, medicine (remember the "four humors"?), chemistry, astrology. The four elements warped and wefted through almost every imaginable aspect of that centered, perfectly balanced Earth which Copernicus would help to knock forever awry. For instance, each of the four seasons partook of its own element: Autumn's dry coldness was earth, winter's wet coldness was water, spring's wet warmth was air and so on. This bore physiological implications of perfect consonance; for instance: "Blood, which increases in the spring, is moist and warm." And because the universe was harmonious, even its superlunary spheres could be described in those terms. Thus the famous astrologer Albertus Magnus could assure us that Saturn was banefully cold and dry while Jupiter was mercifully cold and wet . . .

Do you want to brew a perfume astrologically appropriate to Venus? Well, then, "take musk, ambergris, lignum aloes, red roses, and red coral, and make them up with sparrow's brains and pigeon's blood." The briefest perusal of these ingredients serves up a plausible explication: Venus the planet is consonant with Venus the love-goddess. (Of *course* her perfection is willed; of *course* she engages in perfect circles around us!) And what could be more consonant with romance than blood-red entities mashed up with delicious smells? Just as Venus rules aspects of us on Earth, so we can influence aspects of her with whichever Venusian substances Earth offers.

The more we peer into the pre-Copernican universe, the more harmonies we find. It is both our gain and our loss that those now seem ludicrous.

Earth's appropriate position

Back to motion. We Newtonians and post-Newtonians believe in inertia of rest and inertia of motion. The ancients believed in an *inertia of appropriate position.* In his cosmology, Parmenides had good reason to enclose our world within a sphere of fire; and Anaxagoras had equal cause to postulate that outer space is an ether comprised of heat—for doesn't fire rise above its rival three elements? Doesn't air get thinner on mountaintops? (Sea level atmospheric pressure: 2116 pounds per square foot. At 10,000 feet, the same volume of air exerts a force of only 1455 pounds; at 100,000 feet, 22.6 pounds; at 200,000 feet, 0.44 pound. How can anyone dispute that air clings to its own position above earth and water, then comes to an end?)

By this logic, the sun must be comprised of fire: Sunlight warms us; moreover, the sun moves through the sky, the high place toward which earthly fire reaches. Therefore, the other celestial bodies must also be made of fire. And the appropriate position of fire is above everything else.

Aristotle accordingly writes that each of the entities which exists by nature (as opposed to by human artifice) "has *within itself*" (his italics) "a principle of motion and of stationariness (in respect of place, or of growth and decrease, or by way of alteration." Nineteen centuries later—nineteen centuries!— Copernicus obediently recapitulates that "water, which by its nature flows, always seeks lower places"; moreover, that "earth is the heaviest element; and all things of any weight are borne toward it and strive to move towards the very center of it." This is the logic by which Aristotle argues that Earth, being composed of earth, must be at that element's natural place, the center of the universe.

Hence one of Copernicus's attackers, a certain learned Dominican named Giovanni Maria Tolosani, saves the universe as follows:

Since Copernicus does not understand physics and dialectics, it is not surprising if he is mistaken in this opinion . . .

For Copernicus puts the indestructible sun in a place subject to destruction. And since Fire naturally tends upward, it cannot, except through constraint, remain down near the center as its natural place, as the Pythagoreans falsely hold.

Natural *versus* compulsory motion

Now, what *is* motion? Aristotle subdivides it, seemingly as we do, into "what exists in a state of fulfillment only," what exists only potentially, and what exists as partly potential and partly fulfilled. Indeed, we still use the term "potential energy" to describe the force which resides in a weight of a given mass at a given height, or, as my old physics textbook more accurately and less parochially puts it, "the energy that a system possesses because of its configuration." But we see energy, and matter, too, in more universalist terms than Aristotle did. We can compare the motion of air to that of water, using the same terms for these two antithetical elements: mass, position, velocity, acceleration, distance, direction, friction. One of the most useful tools we now possess for making that comparison is

NEWTON'S SECOND LAW:
FORCE EQUALS MASS TIMES ACCELERATION.

which leads to the very bizarre, but nonetheless true, conclusion that *the mass of any physical object whatsoever is the force on that object divided by its acceleration.* For example, one kilogram equals one newton of force per meter per second per second. (A newton is 3.6 ounces.) When we add to this peculiarity the even more peculiar fact of the existence of a gravitational constant, a few simple algebraic transformations allow us to determine the mass of the Earth, for instance, without needing to weigh it on a celestial balance: the radius of the moon's orbit and the duration of the lunar period (the time the moon takes to complete a single orbit) are all we require! We can compute the mass of the sun just as well from the period and orbital radius of Jupiter as from the period and orbital radius of Venus. "The mind shudders."

But this principle remains unknown to Aristotle, and even to Copernicus. The former firmly states that "there is no such thing as motion *over and above* the things" being moved. The motion of water is fundamentally different from the motion of a star. "It is always with respect to substance or to quantity or to quality or to place that what changes changes."

Having thus claimed that each element has its own natural law, its own "virtue," Aristotle very logically proceeds to the conclusion that the motion of a given body may be (a) natural to it; (b) compulsory, that is, against its nature; or (c) a mixture of the two. For instance, the motion of water being raised in a bucket is compulsory, in opposition to its natural tendency to flow downhill. There is no *natural* Aristotelian motion by which water flows uphill, and any geometer, astronomer or other dreamer who posited a situation in which it did (for instance: evaporation) would have lost the battle in advance. And so Father Tolosani, whom we will meet again in a more menacing context, savages Copernicus as follows:

> A simple body cannot have two natural motions opposed to each other. For we see the Earth moving naturally to the center on account of its natural heaviness. But if it is said to

rotate, its circular motion will be coerced, not natural. Therefore it is false that the Earth rotates with a natural motion . . . thus Copernicus's hypothesis is completely overthrown.

Willed perfection

Aristotle's definition of motion is not only more object-specific than ours; it is simultaneously so much more fundamental that it approaches ontology. It incorporates *agency*. "It is the fulfillment of what is potential *as* potential that is motion." When we say, "I was moved to do this," we are using the terminology of motion in the same way as Aristotle sometimes does. Indeed, in *On the Heavens* we sometimes find Aristotle speaking of the voluntary movement of stars. This is a very ancient view, and very true to our human tendency to see the cosmos as being significant by virtue of being sentient. It is therefore true in the psychological and spiritual senses— or, if you prefer, it is simply beautiful to believe. This is why Ptolemy's equation between mathematics and philosophy can seem so appealing. Before Aristotle, Empedocles supposed Love to be the moving force of the cosmos. Long after him, Dante will end his great poem with a vision of three circles of rainbow and fire; "here power failed the high phantasy; but now my desire and will, like a wheel that spins with even motion, were revolved by the Love that moves the sun and the other stars."

If we posit willed perfection as a fundamental quality of the movements of the superlunary "world," then the mysteries of epicycles become more tolerable still, and the necessity of insisting on their circularity proportionately more urgent.

"Circular movement belongs to wholes and rectilinear to parts"

"A simple body cannot have two natural motions opposed to each other." Why can't it? Because not only would element-specific, place-specific laws of motion be thereby violated, but so would simplicity; so would willed perfection.

"It is impossible that a simple heavenly body should be moved irregularly by a single sphere," accordingly asserts a certain obedient Aristotelian named Copernicus, "for that would have to take place either on account of the inconstancy of the motor virtue . . . or on account of the inequality between it and the moved body." We have already quoted his reaction to either possibility: "The mind shudders."

We begin to see *why* Copernicus's mind might have shuddered. Since he thinks of himself as a geometric rationalist, not as a poet or theologian, what can he do about inconstancy of motion? What can he do with the fact that summer and winter are not of equal length?

Voilà! He defines the uniform motion he longs for as *mean motion*. That's his magic trick.

Years before finishing *Revolutions*, he notes that "in their investigations of the moon's path Ptolemy, and before him Hipparchus of Rhodes, divined with keen insight that the revolution of a nonuniform [motion] must have four diametrically opposite points. These are the maximum swiftness and slowness, and the mean and uniform [motion] at both ends [of the diameter] intersecting at right angles [the diameter connecting] both maxima."

Our savior, Aristotle, showed us how to break motion down into parts. To be specific, he reasoned out what we would now call vectors of motion; but for him motion in different directions partook of different characters: rectilinear and circular motion. The four elements express their inclinations rectilinearly, which is to say upward in the case of fire and air, or downward in the case of water and earth; the celestial bodies for their part inscribe eternal circles on the heavens— one more reason to believe that they are different in their nature from the substances we know on Earth.

Copernicus more or less follows Aristotle in this respect, concluding that "circular movement always goes on regularly, for it has unfailing cause"—hence his refusal to entertain the possibility that the orbits of the planets might be (as they actually are) imperfect circles. Insist on mean motion about mean points; thus runs Copernicus's strategy. The inconveniences which Ptolemy has explained away with equants, we can explain away with rectilinear impulsions if we have to; but with luck we'll never have to. Suffice it to say, as *Revolutions* will, that "circular movement belongs to wholes and rectilinear to parts" which have been disturbed from their proper place. If a celestial orb apparently deviates from circularity, then work out its mean motion and construct enough perfect circles around circles to express that equivalence.

In this book I will tiptoe around the ugly convolutions of Copernican motion theory wherever I can; for I sometimes suspect that even the author of *Revolutions* knows that reducing the heavenly revolutions to uniform circular motion is Sisyphean. Refuting some of Ptolemy's notions on the movements of the planets, he expresses the sadly unambitious wish that "the principles of this art might be preserved, and the ratio of apparent regular motion rendered" not perfectly constant, as one might have hoped, but merely "more constant."

—Well, why not? Regarding Copernicus's definition of uniform motion as mean motion, the astronomer who has read this manuscript to save me from my errors remarks: "Note that we do much the same even today. It's convenient for us to have something called a 'day' that is exactly 24 hours long, but in fact the sun doesn't come back to the same place in the sky at the same amount of time every day."

That's all true. But we know that we're literally cutting corners. Copernicanism for its part is, almost above all, a nobly doomed striving to express planetary orbits as *direct uniform motion on a circular epicycle.*

Stillness

Ever since Newton, we uncentered ones have come to see the importance of *frame of reference* in describing motion. Indeed, for Newton, Galileo and us, constant velocity, not rest, makes our context. Science informs us that Earth is rotating at almost twenty-four hours per spin while traveling around the sun in an orbit whose period is slightly more than 365 days, all the while, along with the rest of the Milky Way, moving toward the Andromeda Galaxy—never mind the momentum from the Big Bang. Let's consider smaller bangs: If we shoot a pistol at a bull's-eye, it may well be that the only motion of interest to us at that particular moment is the bullet's. We would assume the following as a convenience: The celestial motions of shooter, bullet and target can be canceled out when we are considering the bullet's positional change from the chamber into the mound of lead-colored dirt behind the target. Aristotle, however, enshrines convenience into fact when he says:

The innermost motionless boundary of what contains is place. This explains why the middle of the heaven and the surface which faces us of the rotating system are said to be "up" and "down" in the strict and fullest sense for all men; for the one is always at rest; while the inner side of the rotating body remains always coincident with itself.

Before Copernicus, Earth was still. Disregard those dozen-odd impious heliocentrists who asserted otherwise; we ourselves are occasionally afflicted by the presence of Communists and child molesters. Earth was still. *The one is always at rest.*

Let the following definition be a parable for the whole centered Earth and why it took so long to dislodge. This comes from Plato's *Timaeus*: *Heaviness is a body's resistance to being moved from the place where it belongs.*

Frame of reference was also well understood by Copernicus, who now begins his famous argument with the remark that "every apparent change in place occurs on account of the movement either of the thing seen or of the spectator, or on account of the necessarily unequal movement of both." In other words, place does *not* have to be an innermost motionless boundary. We see the sun rise and set; we see the heavens appear to revolve. Therefore, one can entertain the supposition that it's the heavens which move, or one can suppose (perish the thought!) that it's our Earth which moves in the opposite direction from the perceived celestial turning.

I.5: "Does the Earth have a circular movement?"

Here *Revolutions* commits itself to the heliocentric hypothesis, and so unequivocally that the account of its dying author's fury at Osiander's preface violates no psychological "appearances." The statement of commitment deserves to be quoted, not because it has been put especially elegantly, and certainly not for any pretense of originality—on the contrary, Copernicus hastens to strengthen his case by invoking three Pythagoreans and a Syracusan—but because it memorializes a man's public adherence to a true idea. Like the "I dos" of a bride and groom, what matters is not only how the thing was said, but also *that* it was said, and in the hearing of a world whose center may well be eccentric to the lovers' hearts:

> For the daily revolution appears to carry the whole uni-verse along, with the exception of the Earth and the things around it. And if you admit that the heavens possess none of this movement but that the Earth turns from west to east, you will find—if you make a serious examination—that as regards the apparent rising and setting of the sun, moon, and stars *the case is so.*

(Osiander: "It is not necessary that these hypotheses should be true, or even probable; but it is enough if they pro-vide a calculus which fits the observations . . .")

Furthermore, Copernicus continues, "the fact that the wan-dering stars," meaning the planets, "are seen to be sometimes nearer the Earth and at other times farther away necessarily argues that the center of the Earth is not the center of their cir-cles." (Ptolemy never said that it was, exactly. He said that it was the center of whatever deferents, epicycles and equants might be needed.) *The case is so* that the Earth not only rotates on its axis, which would explain day and night, but also expresses a secondary motion in relation to those to-and-fro-ing planets—namely, the yearly revolution along the ecliptic—although at this stage of the argument, Copernicus has not ruled out the contrary possibility that that secondary motion may be intrinsic to those planets rather than to the Earth.

The axe has fallen now; the old universe has received a gaping wound.

I.6: The geometry of heavenly immensity

"And God made the firmament," or dome, we read in Genesis, "and separated the waters which were under the firmament from the waters which were above the firmament. And it was so. And God called the firmament Heaven." As for Copernicus, he now parallels God's great construction as follows: "Now let the horizon be the circle *ABCD*, and let the Earth . . . be *E*, the center of the horizon" which separates the visible from the invisible stars.

In our situation of Copernicus's universe between Augustine's and Herschel's, the following demonstration was already referred to. It may be appropriate to reproduce its full logical train here, so that one can get a sense of how *Revolutions* reasons. Most of the book is so murkily written, its terms so undefined, its half-correct generalities so riddled with special cases (for an instance, see my exegesis of the very last chapter), that any brief summary of the process from observation to conclusion remains practically impossible. This argument, whose analogue appears in *The Almagest*, is a successful exception.

Imagine *E* as a tiny sphere, on which stand you and I, complaining about Ptolemy and observing constellations through the instrument called a dioptra. Let *C* be the point on the horizon where the first star of Cancer rises. Let *A* be the point where the first star of Capricorn is setting at the same time. "Therefore, since *AEC* is in a straight line with the dioptra, it is clear that this line is a diameter of the ecliptic, because the six signs" of the Zodiac between Cancer and Capricorn inclusive "bound a semicircle, whose center *E* is the same as the horizon." When one revolution has occurred—it matters not whether of the celestial sphere or of *E* itself, but Copernicus with typical inconsistency must mean *half* a revolution, or one complete transit across the visible horizon—and Capricorn begins to rise at *B*, then Cancer will set at *D*, in which case "*BED* will be a straight line and a diameter of the ecliptic." Since *AEC* is also a diameter of the ecliptic, and since *E* is the center of both *AEC* and *BED*, "the horizon always bisects the ecliptic." (In fact Copernicus has not proven his "always" with merely two cases, but the claim is highly believable.) Therefore, since by the rules of geometry a circle which bisects a great circle must be a great circle, both the horizon and the ecliptic are great circles of the celestial sphere. "Therefore the horizon is a great circle and its center," which passes through the center of the Earth, "is the same as that of the ecliptic," whose center is near the center of the sun.

"From this argument it is certainly clear enough the heavens are immense in comparison with the Earth," sums up Copernicus—and even in our own persnickety epoch, we rarely care whether we measure astral distances from the center of the Earth or from the observatory dome. Objects within the solar system are another matter, but only for astronomers: Mariners need not apply parallax corrections when sighting on any planets, no matter that their distance from us varies considerably, the overall *vastness* of the heavens being so great that no practical harm can be done by envisioning both

Mercury (if we're lucky and skilled enough to see it) and Saturn as lying upon the "ceiling" of the same celestial sphere.

Copernicus continues: "But we see that nothing more than that has been shown, and it does not follow that the Earth must rest at the center of the world."

Then where *is* the center? Since the "wandering stars" do not always maintain the same distance from the Earth, "it is necessary that movement around the center should be taken more generally, and it should be enough if each movement is in accord with its own center."

I.7–9: Copernicus almost defines gravity

Since Earth is made primarily of earth, one of the two heavy elements, all elements which contain weight are inclined to fall down toward the center of the Earth and stay there. "All the more then will the Earth be at rest at the center." That is how Ptolemy and the other Aristotelians put their case.

"I myself think," responds that good Platonist Copernicus, "that gravity or heaviness is nothing except a certain natural inappetency implanted in the parts by the divine providence of the universal Artisan, in order that they should unite with one another in their oneness and wholeness and come together in the form of a globe." And in such a case, why shouldn't the other planets and even the sun also partake of this quality?

A digression on Neptune's atmosphere

It is worth reiterating that this is still not quite gravity as we understand it.

Addressing one of Ptolemy's objections to the idea of a rotating Earth, that surely everything and everybody on Earth would be flung off and scattered into space were we not at rest, Copernicus chucklingly demands why his forbear didn't suffer anxiety over the fate of the cosmos instead? By Ptolemy's own logic, isn't it revolving around the Earth with incredible velocity? Moreover, the farther away each sphere's position, the more rapidly it must turn to complete the same revolution. Why don't the inhabitants of the Sphere of Fixed Stars fly apart into pieces? (My physics textbook informs me that "small steel balls of radius 10^{-4} meter, rotated at . . . high frequencies, explode when the peripheral speed attains approximately 1000 meters/sec[ond].") But very few stars have exploded from speeding as of yet. Stars aren't steel balls, to be sure; whatever perfect superlunary element forms them may be proof against extreme centrifugal force; all the same, positing an unknown substance with unknown properties remains an unconvincing defense against a real difficulty. Ptolemy's argument for immobility bears less conviction than Copernicus's whirling objection.

We uncentered individuals know that we do not fly off our spinning planet because gravity and inertia, with centrifugal complications, keep us here. We further know that our atmosphere is matter, therefore possesses mass and accordingly experiences inertia and gravitational attraction just as we do, according to Newton's laws. But Ptolemy respected Aristotle's laws. For him, air's motion had nothing to do with earth's, for remember the principle that each element contains "*within itself* a principle of motion and of stationariness." Moreover, the (seemingly) stationary atmosphere comprised another case of his "scattered into space" argument: When an arrow rushes through the air, the air declines to move with it. Therefore, if the Earth did what an arrow does, the atmosphere would likewise refrain from motion, and we would see clouds and meteors (the latter were then believed to be atmospheric phenomena) continually moving away from us in the opposite direction from our rotation.

What would he have made of Neptune's case? This planet revolves from west to east, as the Earth does. But—at the equator, at least—the Neptunian winds blow east to west at two thousand kilometers per hour, the greatest velocity of any atmosphere in the solar system. Had Ptolemy been incarnated as an equatorial Neptunian, deduction and observation might have convinced him—for utterly spurious reasons—that his home planet did rotate just as Copernicus claimed.

A sub-digression on the Coriolis Effect

As it happens, Ptolemy's objection was more to the point than both he and Copernicus knew. Just as the Sphere of the Stars, if it actually existed, would revolve westward around us more rapidly than any nearer sphere, so the Earth's equator, being by definition the widest part, revolves eastward faster than the rest of our globe; while the poles remain (theoretically) still. Wind currents passing due north from the equator accordingly carry with them an eastward momentum of rotation greater than that of the land or sea beneath. Accordingly, our north-directed wind becomes a northeasterly. Conversely, a wind blowing due south toward the equator finds itself traveling with ever less speed in proportion to the terrestrial surface below, and therefore takes on a southwesterly motion with respect to that surface.

In the southern hemisphere, this effect (named, as you might have guessed, after a certain Coriolis) gets reversed: South-blowing winds go southeastward; northerly winds go northwestward.

The Coriolis Effect exerts itself in myriad ways. For instance, the vortices of cyclones turn counterclockwise in the northern hemisphere, clockwise in the southern. It applies to water as well as to air, and accordingly influences the paths of major ocean currents.

Why didn't Ptolemy and Copernicus recognize it? First of all, they lacked what we would now refer to as data histories of air and water flows for the northern and southern hemispheres. Secondly, the Coriolis Effect is moderated, and at times utterly overruled, by local pressure forces and frictional variations.

In brief, our rotating Earth *does* leave its atmosphere behind, but neither as completely nor as violently as Ptolemy imagined.

"Then what should we say about the clouds?"

At any rate, Copernicus could have scotched Ptolemy's argument about the atmosphere left behind had Newtonian mechanics been available to him. Particles and worlds do "unite with one another in their oneness and wholeness." Once gravity's universality is understood, how simple it all becomes! Herschel explains the matter to our satisfaction with an analogy which would have been available even to Ptolemy: On an ocean cruise, a ball thrown straight up lands back in the thrower's hand, because thrower and ball are imbued with the same motion from the ship. Copernicus's refutation of Ptolemy could almost have been Herschel's; but he felt obliged to couch the refutation thus:

> Then what should we say about the clouds [not continually speeding westward against an eastward terrestrial rotation] . . . except that not only the Earth and the watery elemeny with which it is conjoined are moved this way but also no small part of the air and whatever other things have a similar kinship with the Earth, whether because the neighboring air, which is mixed with earthly and watery matter, obeys the same nature as the Earth, or because the movement of the air is an acquired one, in which it participates without resistance on account of the contiguity and perpetual rotation of the Earth?

In other words, Copernicus falls back rather tentatively on two explanations, first, the old elemental one that terrestrial air's fundamental lightness has been infected by a degree of heaviness by contact with earth and water, and second, an intuitional expression of inertia. To me there is something very moving about this intellect which never finds the tools which Newton found but which gropes and reasons its way just the same toward an accurate apprehension of reality.

I.9: Centering the sun

The most spectacular example of this in *Revolutions* now comes suddenly, when after preliminaries which suggest only that it's plausible to suppose that the Earth moves in some fashion, Copernicus takes the bull by the horns and says it:

"Lastly, the sun will be regarded as occupying the center of the world. And the ratio of order in which these bodies succeed one another and the harmony of the whole world teaches us their truth, if only—as they say—we would look at the thing with both eyes."

The Limits of Observation in 1543

How easy it used to be to save the appearances!

I f only we would look at the thing with both eyes! But most eyes cannot distinguish stars which happen to be separated by less than four arc-minutes.

The less we know, the freer we remain to conceptualize. (Orwell: "Ignorance is strength.") An accuracy-loving science-fiction writer of the 1950s was at liberty to imagine Venus as a world of swamps and jungles; we didn't know how hot it was. Remaining unaware of the existence of Neptune, let alone of the speed of that planet's equatorial winds, Ptolemy remained free to arrive at an incorrect conclusion about the Earth's atmospheric rotation based on irrelevant observational evidence. A savant in the 1550s was equally unfettered in being able to choose Copernican heliocentrism or Ptolemaic geocentrism. Either system went far to "save the appearances." Indeed, Copernicus's various failures and incompletenesses, not to mention that reassuringly modest preface imposed upon him by Osiander ("let us permit these new hypotheses to make a public appearance among old ones which are themselves no more probable"), obscured much of the threat Copernicanism might someday pose to a literalist Scriptural universe: Why not believe in heliocentrism and a Throne of God at the same time?

Foucault's pendulum

What is reality? The history of science, not to mention life itself, teaches us to suspect that more will always exist than we have yet apprehended. Ptolemaism is a parable of our way of being: It suits us; it explains almost everything; let's ignore the rest; we'll save *all* the appearances someday! Then comes Copernicus; and after him, Kepler, Newton, Einstein, accompanied by rocketeers who aim their lenses at our moving Earth! ("The rocket photographs show that the most striking impression of an outside observer would be the flatness of the Earth's surface . . . The highest mountain peak and deepest canyon would be but minor irregularities on the relatively smooth surface of the Earth." From a sufficient altitude, the sublunary realm becomes superlunary. Thus the limits of observation.)

Ptolemy insists that if (perish the thought) our Earth actually revolved, falling bodies would strike the ground in a place behind the actual perpendicular. Ptolemy is correct, within the limits of observation. It staggers me that not until 3 February 1851 will Foucault's pendulum extend those limits, which it does as follows:

By means of a wire, suspend a weight from an immobile tripod. Loop a thread around the wire, slowly pull the thread tightly due southward—that is, along the Earth's axis of rotation—until the wire gets drawn toward you, be patient for a moment or two while any tremors in the wire die out, then touch a match to the thread until it parts. The pendulum thus begins a series of swings uncontaminated by any lateral motion. What happens next recapitulates the Coriolis Effect: If you happen to be north of the equator, each northern extremity of the swing will be eastward of the previous one; each southward extremity will be successively more westward. (In the southern hemisphere the swings go counterclockwise.) Foucault's pendulum is falling freely in space, while the Earth revolves around it! And now for an inversion of the Coriolis Effect: This motion will be most pronounced at the poles and utterly absent at the equator. It is represented mathematically for any given location as fifteen degrees per hour (the rate of rotation of the Earth) times the sine of the latitude.

Go to Paris, and in the Pantheon, beneath the domes windowed and windowless, you will see a replica of Foucault's pendulum slowly swinging to and fro within its graded ring, silent, its alteration almost imperceptible. From a distance and at the appropriate angle, the gleaming ball seems hardly to move, only to pulse very slightly. Moreover, it pauses for a moment at the end of each swing, which it then retraces, although not quite; in Paris the path requires about six minutes to change by one degree. That comes to 11 degrees per hour (15°/hr × [sin [lat 48° 51'] = 15 × 0.7528 = 11.292), 272 degrees per day. YOU ARE INVITED TO COME SEE THE EARTH TURN, proclaims the sign; but it is astonishing how long one must actually watch before becoming certain of the clockwise alteration of the swing. Silently, subtly, the Earth turns (1674 kilometers per hour), observed by frozen-breasted statues. If you watched only one or two swings, you might never know that it moves—another example of the limits of observation. If you witnessed the Coriolis Effect without completely understanding it, you might predict the most rapid rotation of the pendulum at the equator, not the poles. Sometimes, as in Ptolemy's case, the limits of observation get maintained by false interpretations of previously true observations.

Reality is what we perceive now. What a pathetic, parochial definition! But it is the truth. Thanks to the transmissions of our orbital telescopes and interplanetary probes, the planets, at least, have begun to reveal trifles of their glory: the reddish craters of Mercury swarming in space's blackness (I've read that Copernicus never observed Mercury even as a dot of brightness; he laments the obscuring vapors of the Vistula in this regard; but indeed, my current edition of *Norton's Star Atlas* assures me that "the general experience" of searching for Mercury even with an amateur telescope "is one of frustration and disappointment"), the blue and ocher orb of Mars, which, thanks to iron, is actually redder than it appears in many photographs, "partly a result of illumination conditions, and partly due to calibration difficulties in the observing system"—reality-as-perception again!—the spectacular, jellyfish-glowing disk of Jupiter, whose moons, utterly unknown to Copernicus, orbit round it like marbles of semiprecious stone; the black crater, named after him, which gazes forth like an eye from the sea of debris on our moon's grey horizon.

"Bequeathed like a legacy"

Our heliocentrist hero's only possessions in the way of perceptual wealth are (1) the pear-shaped, clocklike, symbol-reliefed disk of Martin Bylica's astrolabe, which Copernicus seems to have used in 1494; (2) the torquetum, which resembles a globe upon steeply swiveling metal sheets, a quadrant which he first recommends that we construct of wood, then thinks better of that, because wood might warp and mislead us; (3) his eyes, which are less than superhuman and can therefore distinguish five planets at best; and—(4) most precious of all, it would seem, the observations of the ancients, including Ptolemy, the father whom he's growing beyond but never leaves behind. He knows, for instance, that "Saturn, the highest of the wandering planets, completes its revolution in thirty years, and the moon which is without doubt the closest to the Earth completes its circuit in a month." It is from Ptolemy and that crew that he knows it. *Revolutions* brims with such citations as "Ptolemy marked the greatest southern latitude as being approximately 7° in the case of the perigee of Mars . . ."

He avoids cribbing uncritically from Ptolemy. *The Almagest* gives Saturn's apogee as 224° 10'. Copernicus puts it at 240° 21' in Book V, Part 6; at 226° 30' in Book II, Part 14. All the same, without Ptolemy and the torquetum, where would Copernicus be? (Indeed, his instruments are the same as Ptolemy's.) He does the best he possibly can; he aims for success within the limits of observation. Or, as a capsule biography puts it: "He made some observations, but they did not help him since he used clumsy instruments made by himself. So he turned to old, faulty observations." These latter were not all as faulty as the biography implies—for instance, Ptolemy's value for the maximum angular elongation of Venus scarcely differs from our own—all the same, they sometimes misled him as much as his fidelity to uniform circular motion. For instance, many ancient measurements of precession were wrong. Accordingly, Copernicus's theory of precession will be wrong.

Do you know what Ptolemy called those old, faulty observations? "The work for another's love of wisdom and truth." He was right. Knowing and respecting this, Copernicus advises us to "hold fast to their observations, bequeathed like a legacy." Truth, at least of the scientific kind, is arrived at (approximated, I should say) only in many increments of ghastly drudgery.

Tycho Brahe (who incidentally cannot bear the Copernican heliocentric hypothesis, and in his day a hypothesis is all that it remains, thanks to limits of observation) will expend half a dozen years and nine thousand folio pages of crabbed writing to learn that the actual position of Mars sometimes varies by a full eight minutes of arc from where it should be according to his own theories. Kepler will then courageously decide that "since it was not permissible to ignore them, those eight minutes point the road to a complete reformation of astronomy." Bequeathed like a legacy! Copernicus, you will recall, had stated that an accuracy of ten *degrees* of arc (one degree equals sixty minutes) would have made him "as elated as Pythagoras must have been after he discovered his famous principle." (*Revolutions*, Book VI: "But neither 3 minutes nor 4 minutes is large enough to be measured by means of an astrolabe; therefore that which was considered to be the greatest latitude of obliquation of the planet Venus is *correct.*") By the ten-degree standard, the eight minutes would never have tortured Brahe and Kepler; the reformation would never have occurred.

Oh, the limits of observation! Here's how Kepler described Copernicus Crater and its fellows: "The dark spots on the moon are, therefore, a kind of liquid which, by its coloring and softness, blunts the light of the sun." No observation could yet disprove the hypothesis that lunar craters are filled with liquid.

And by the way, why did Tycho reject Copernicanism? Because this skilled and dedicated observer couldn't detect the annual parallactic motion of stars, which Copernicus's theory said should be visible if the Earth moved. (We'll get to those in two chapters.) Tycho therefore reasoned out his own immobile Earth accompanied by an orbiting sun about which the planets circled. He saved the appearances as well as Copernicus. Parallax was not scientifically validated until 1838.

"Binoculars are usually needed"

What then can Copernicus hope to accomplish from that unassuming brick structure, rectangular in cross-section, which rises up to a tiled arrowpoint and has now become known as the "Copernicus Tower"? What can he actually see above Frombork's spire-needles in the evening light? Well, in addition to the stars there are certain peculiar points of light which he informs us "wander in various ways, straying sometimes towards the south, and at other times towards the north—whence they are called 'planets.' " An amateur sky-watching guide from my own era advises us to find them by "watching their movement from night to night against the backdrop of seemingly fixed stars, noting how they disturb the shapes of the constellations." From night to night! That doesn't necessarily sound easy. Venus will be recognized as such because it is brighter than other celestial bodies; Jupiter is almost as brilliant, but sometimes Mars, which resembles an orange star, is more so; Saturn seems to be a yellowish star; as for Mercury, "binoculars are usually needed to spot it," which must be why Copernicus never recorded any observations about Mercury.

The limits of observation half-crippled him. You or I can look at moon rocks. Copernicus had to rely on treatises hasped and studded with metal decorations and inhabited by a half-grown spherical geometry some of whose axioms he needed to work out for himself, in order to calculate how far away the moon was; meanwhile, this leg or that angle of the triangle which his genius projected upward into the Planetary Spheres could only be known through the observations of dead astronomers and astrologers whose theories he couldn't trust. (Do you remember the Alphonsine Tables? Copernicus's library was afflicted with that albatross.) Ironically, his misplaced faith in those frequently erroneous observations introduced so many contradictions as to encourage him all the more to reject Ptolemy's system, which couldn't explain wrong planetary positions any better than right ones.

"We are approaching the limits of our ability to probe the skies." This sentence appears in an astronomy textbook published in 1982. "The subsequent 23 years of astronomical discoveries have proved it to be wildly inaccurate," another astronomer tartly responds. "Observational capabilities have expanded enormously since then." But in 5382, if our species continues to exist then, we will still in our finitude be grappling with the limits of observation. What then to do but deduce and infer?

In 1543, Copernicus similarly approaches the limits of observation: the five planets are but points of light, and he's only seen four of them! Steadfastly, he—deduces and infers. "And as far as hypotheses go," snickers Osiander in that patronizing preface, "let no one expect anything in the way of certainty from astronomy . . . lest, if anyone take as true that which has been constructed for another use, he go away from this discipline a bigger fool than when he came to it. Farewell."

Meanwhile, without certainty or even sight (he'll never see a planetary disk), lacking Foucault's pendulum, taking as true that which has been constructed for another use, with few facts, without conclusive proof, Copernicus uncenters the Earth! "Thus the heliocentric system appeared almost two centuries too early to be properly understood and appreciated." All the same, in spite of the limits of observation, it did appear. And that's why I call Copernicus a great man.

Exegesis

I.10–14

A nd so the sun is the center of our "world." How might the other heavenly bodies be arranged?

I.10: Simplifying and rearranging the heavenly spheres

Copernicus's precursors reasonably argued that the heavenly orbs which seemed to cross the sky more slowly must be farther away from us than those which move rapidly, such as the moon. More precisely, "the magnitude of the orbital circles shall be measured by the magnitude of time." They thus correctly ranked them, in increasing order of closeness: Saturn, Jupiter, Mars.

As for Venus and Mars, the ancients differed as to whether they were above or below the Sphere of the Sun. Because Venus's orbit is a Copernican paradigm of sorts, we will discuss it in the next chapter. For now, let's simply note that Copernicus schematizes the "spheres" of the inner and outer planets rather well:

It is necessary that the space left between the convex orbital circle of Venus and the concave orbital circle of Mars should be viewed as an orbital circle or sphere homocentric with them in respect to both surfaces, and that it should receive the Earth and its satellite the moon and whatever is contained beneath the lunar globe.

This logical deduction from heliocentric premises Copernicus is not averse to presenting as a cause:

"Therefore we are not ashamed to maintain that this totality—which the moon embraces—and the center of the Earth too traverse that great orbital circle among the other wandering stars in an annual revolution around the sun."

Revolutions's translator remarks here that "Copernicus has telescoped the eccentric circle of Venus and that of Mercury into one circle carrying the Earth; and he has furthermore collapsed the three epicycles of Saturn, Jupiter and Mars into this same one circle. That is to say, one circle is now doing the work of five."

"I also say," repeats Copernicus in yet another contradiction of Osiander's preface, "that the sun remains forever immobile and that whatever apparent movement belongs to it can be verified of the mobility of the Earth." Furthermore, he says, bravely measuring the depths, the distance from the Earth to the sun, vast as it is, is nothing in comparison to the distance from the Earth to the Sphere of Fixed Stars.

I.11: The Earth's three movements

And now Copernicus replaces Ptolemy's two heavenly movements with three of his own.

Firstly, the Earth rotates from west to east each day, which "describes the equator or equinoctial circle."

Secondly, the Earth's annual circle about the sun also moves from west to east. This circle lies between the circles of Venus and of Mars. "So it happens that the sun itself seems to traverse the ecliptic with a similar movement . . . when the center of the Earth is traversing Capricorn, the sun seems to be crossing Cancer; and when Aquarius, Leo, and so on . . ."

Copernicus postulates a third Earthly movement westwards, from Aries to Pisces. This movement, called declination, seems necessary to explain the fact that the Earth does not maintain the same position with respect to the plane of the ecliptic. Copernicus asserts the necessity for this because on most parts of our globe, the ratio between day and night alters throughout the year, and seasons succeed each other.

Declination and orbital rotation are opposite in direction but almost equal, he opines. "It follows that the axis of the Earth and the greatest of the parallel circles on it, the equator, always look toward approximately the same quarter of the world"—approximately, not exactly, since precession can't be denied; the equinoxes and solstices have altered by twenty degrees since Ptolemy's era.

Copernicus sums it up: "The ecliptic remains perpetually unchangeable—the constant latitude of the fixed stars bears witness to that—while the equator moves." He is right about this; before him, astronomers had thought that it went the other way.

I.12–14: Some theorems of plane and spherical geometry

Book I now ends in an explosion of geometrical proofs: arcs and chords, subtentions, inscribed quadrilaterals, plane rectilinear triangles, spherical triangles. "Since we see that we have come so far that the difference between the straight and the circular line evades sense-perception as completely as if there were only one line . . ." All this, and especially that lengthy collection of numbers in his "Table of the Chords in a Circle," I propose to spare you, with this justification from Copernicus himself: "For if they had to be treated in greater detail, the work would be of unusual size."

Orbits of Venus

Where were we? At the center of the universe, of course. Now, what about Venus?

That orb, which to the Mesopotamians was fertile Ishtar whose talismanic image, when carved out of lapis lazuli at the hour when Venus ascends into Taurus, helped men to win the hearts of women, while to Copernicus and Ptolemy it was a wandering star and to us is "a veritable inferno of shadowless semi-twilight, filled with hot carbon dioxide under enormous pressure," gets dealt with piecemeal in *Revolutions*. In this sad little tract of mine—incomplete crib of an unreadable, error-ridden soliloquy addressed to a future which its author might have fled—I cannot hope to do justice to many of Copernicus's mathematical narratives; nor do I want to. Ignore Venus's retrograde rotation—a feature shared, so far as we know, only by Uranus among all the planets in our solar system. Never mind that from one **inferior conjunction** to the next (584 days), or from **superior conjunction** to superior, Venus apparently circles us Earthlings four times when in post-Copernican fact it circles the sun *five* times. The reason has to do with certain coincidences having to do with the angular velocity of the two planets' solar rotations.

(Do you care to know what conjunctions are? In the pre-Copernican universe, they occurred whenever two planets passed through the same sign of the Zodiac. This could be astrologically favorable or not, depending on the prior sympathy or contradiction between the orbs in question. "Now all planets are afraid of a conjunction of the Sun," claims one occultist, "rejoicing in the trine, and sextile aspect thereof." For our modern uncentered definition of conjunction, see below, this chapter.)

"In the fifth place Venus is carried around in nine months," asserts Copernicus, who's estimated a month and a half too long, for one solar revolution of Venus actually takes 224.7 days. Never mind that, either. He gets some things right.

Copernicus imagines the Venusian orbit as comprising (what else?) a circle eccentric to the revolution of our own planet. He determines that the radius of the Venusian orbital circle is 7193 compared to the ecliptic radius of 10,000. We now calculate the mean radius of the Venusian orbit as 0.7233 times that of our own, and if we divide 7193 into 10,000 we derive the remarkably close match of 0.7193.

His orbit of Venus is relatively uncomplicated; and Jacobsen remarks: "All motions in the diagram are direct . . . Because of the small value of the eccentricity of the center and the moderate value of the inclination of Venus, this arrangement fairly well represents the planet's motion in longitude."

The true orbit of Venus lies even farther beyond my capabilities of description than it excels Ptolemy's and Copernicus's. But it is worth interrupting our exegesis to consider the

Venusian case, however superficially, since Copernicus "seems to feel that he has won his case with an argument from Venus and Mercury."

"In line with the Water-Bearer's testicles"

As we saw in *Revolutions* I.10, the first question he had to solve was: Where *is* Venus?

Both Ptolemy and Copernicus studied that question.

"And in the year 21 of Hadrian, Egyptianwise Mechir 9–10 in the evening, we ourselves observed Venus at its greatest elongation from the sun," Ptolemy reports. "It was very nearly 2/3 a full moon east of the northernmost star of the four in a square, following the star to the east of, and in a line with, the Water Bearer's testicles; and it seemed to outshine the star." From his own sightings and from those of Theo the Mathematician, he determines just where within the ecliptic fall Venus's apogee and perigee of eccentricity: twenty-five degrees within the Bull, and twenty-five degrees within the Scorpion, respectively. Then he draws three overlapping circles and works out the radius of Venus's epicycle and of the deferent around which the epicycle rides.

Meanwhile, in 1529, "on the 4th day before the Ides of March," our reclusive cipher-genius himself observes Venus "1 hour after sunset and at the beginning of the 8th hour after midday. We saw the moon begin to occult Venus at the midpoint of the dark part between the horns, and the occultation lasted til the end of the hour or a little later . . . at the middle of the hour or thereabouts, the centers of the moon and Venus were in conjunction, and we had a full view at Frauenburg," he triumphantly concludes.

Quite so; now what about the circles which both of our heroes have drawn?

Closest of all to our immobile and eternally centered Earth spins the Sphere of the Moon, where eternal perfection begins. We already knew that. Farthest of all, the Sphere of Fixed Stars twirls for all time, or at least until the Last Judgment. That too goes without saying. In between those superlunary bounds, all planets revolve around us, in the following increasing order of nearness, as we know from their decreasing **periods** (a period is the time needed to complete one orbit): the Sphere of Saturn, the Sphere of Jupiter and the Sphere of Mars. Below them, not without doubts as to exactly where, Ptolemy places the Sphere of the Sun. His doubts have to do with where the Spheres of Venus and Mercury should go.

Parallax

The Sphere of Venus must be above the Sphere of the Sun, not below, argue certain ancients, Plato among them. Were it below, which is to say between us and the sun, the sun would sometimes "be eclipsed in proportion to" the magnitude of the Venusian disk, and we've never seen that happen. (Eventually we will, but not until 1639; that nineteenth year of the telescopic era marks our first observation of a transit of Venus across the sun. Since this phenomenon occurs at odd intervals of between 8 and 121 years, we are lucky to have seen the truth so soon. As for Mercury, the Mesopotamian god of wisdom, an astronomer from my own time informs me that that orb is likewise "too small to see without a telescope." In short, Plato stands excused. It's no condemnation of him that transits of Mercury take place commonly—one or two every three years.)

Acknowledging a dearth of reliable observations regarding those two "wandering stars," Ptolemy decides to trust the judgment of contrary ancients, and place them higher than the moon and lower than the sun, although his certainty on the subject remains incomplete because Venus and Mars display no "sensible parallax."

As Copernicus's translator reminds us, there is *always* parallax.

Parallax, being a geometrical beast, is often measured in radians and arc-seconds. It possesses its own wearisome kinds and categories; suffice them to be explained as follows:

Nowadays, when we measure the position of a star, we take two bearings on it, the second one six months away from the first, so that Earth has traveled as far as possible away from its original position. The location of a nearer orb such as the moon can be calculated from two positions on Earth itself, meaning either two points in space, or one place at two different moments of Earth's spin through the night. In either case, triangulation, one of the fundamental principles of terrestrial as well as celestial orienteering, requires us to treat the object under consideration as one point of a triangle, then sight in on it from two other points which can be considered the remaining vertices of that triangle.* Those two bearings from two known positions confirm the position of the unknown point.

---

*The "astronomical triangle" for the heavenly body T comprises: (1) the observer's meridian (the great circle from the nearest celestial pole through the observer's **zenith**, or the point on the celestial sphere directly overhead); (2) the half-great "hour circle" from that same celestial pole through T through K, the intersection with the equator, through the other celestial pole; and (3) the vertical circle from the zenith through T. Spherical geometry then allows us to calculate one of these three sides if the other two are known.

As our point of reference moves through its heliocentric orbit, the position of the star whose distance from us we seek accordingly appears to change its angle in relation to its background stars, but—fortunately for Ptolemaic geocentrism— not very much. This is why Ptolemy called stellar parallax insensible. Had Copernicus been able to detect it, *Revolutions* might have suffered fewer vicissitudes. But Copernicus, remember, is the one who said: "If I could bring my computations to agree with the truth to within ten degrees, I should be as elated as Pythagoras." And the truth of parallax lurks within a hiding-hole of much less than a single degree. ("I'll say!" an astronomer writes me. "The largest parallax is 0.7 arcseconds, so less than 1/5000 of a degree.") One of the reasons that Tycho Brahe would ultimately reject the Copernican solar system was that in spite of all his painstaking observations (and he is generally conceded to have been one of the greatest skygazers ever), he never succeeded in measuring the parallax of any star. The reason was precisely what both Ptolemy and Copernicus had previously stated: In regard to the immensity of Heaven, the Earth ought to be considered the merest geometric point. But, as we've seen, that's not mere enough! How can we blame them for drawing back from Herschel's universe? Ptolemy truly writes: "It is clear that in the case of those stars having no sensible parallax (that is, those with respect to which the Earth is in the ratio of a point) getting the ratio of the distance would be impossible."

Needless to say, Ptolemy would not have followed the procedure given above, since he did not believe that the Earth moved. For details, consult his good disciple Copernicus, who informs us how to construct a replica of Ptolemy's parallacticon: Three long straight-edges ruled to at least 1414 parts, pivots and eyepieces allow an observer to measure the distance of a celestial orb from the vertex of the horizon, "and by means of the table," furnished courtesy of *Revolutions*, "he will get the sought arc of the great circle passing through the star and the vertex of the horizon."

The parallax of the moon is relatively easy to measure, that orb being close. Ptolemy computes it as 1° 7', from which he derives an Earth-moon distance at the moment of observation of forty times the terrestrial radius plus twenty-five minutes of arc.*

The parallax of Saturn, outermost of the planets in the Ptolemaic cosmos, remains measurable, no matter that the process is less convenient. But from you and me to, say, our near neighbor, Vega, twenty-five weary **lightyears** go on and on. A lightyear is the distance that light travels in a year. Twenty-five lightyears equal 237,000,000,000,000 kilometers, an incomprehensible number which we will pretend to understand by expressing it in scientific notation as 2.37×10^{14} kilometers. In comparison to this distance, the entire ecliptic circle which Earth traces about the sun may be considered for many

---

*An astronomer comments: "This seems to be mixing apples and oranges, or rather lengths and angles. It seems akin to say 'I am six feet plus three degrees tall.'" The astronomer is correct. This complication, alas, recurs throughout *Revolutions*, and throughout *The Almagest* as well. It took me an embarrassingly long time to grasp that to turn apples into oranges, all I needed to do was divide each measure of arc-minutes or arc-seconds by 60 to obtain the proper ratio: 40.42. Never mind my stupidity. The mean Earth-moon distance is presently calculated at 60.27 times the Earth's radius at the equator, but the 40.42 was not Ptolemy's final number, which seems to be "the straight line *EA* or the mean distance at the szygies," exactly 59 times the Earth radius. This is a closer match.

practical purposes, especially given the crudity of Tycho's instruments, as—a point.

Vega's parallax is about 0.13 arc-second. Pity the technology of Ptolemy's day, which had nearly all it could do to subdivide a brass circle into 360 relatively equal degrees of 60 minutes each. Had Ptolemy made 60 subdivisions more (and he actually did construct his own instrument, with rods and prisms, "dividing the line so defined on the fixed rod into 60 parts and each of these into as many as possible"), he would have gotten down to arc-seconds, but then he would have needed to dissect each arc-second into a hundred more parts to be able to measure Vega's parallax. That works out to 129,600,000 marks in all. What if he had settled for ten parts, and rounded 0.13 arc-second down a nice even 0.1 arc-second? Then he would still have been off by more than seven and a half lightyears!

"While true, this seems an odd statement," writes Dr. Eric Jensen. "If Ptolemy had done this, he would have been astonishingly close to the right answer—the fact that it's a little off doesn't seem that relevant." Yes, but seven and a half lightyears! That's 6.64×10^{13} kilometers' erroneous journey into Herschel's unending and incomprehensible darkness, against which we once upon a time built against our fear a still more erroneous Sphere of Fixed Stars, the final bulwark of everything, which all the same, considered in comparison to that darkness, might as well once more have been the merest point . . .

As for Tycho, his attempt to find parallactic shift would have failed even had he lived nine years longer and beheld Galileo's telescope. For him the nearer stars did not move in relation to the farther; therefore the Earth must not move, either. Stellar parallax remained unobserved until 1838.

The parallaxes of Venus and Mercury seemed to Ptolemy impossible to compute because "they are hidden at the time they are in conjunction with the sun, and they show only the digressions which they make on either side of the sun; hence they are never found without parallax." These words are Copernicus's, not Ptolemy's. Conjunction refers to the state when the Earth, the sun and the planet in question all lie in a straight line. At inferior conjunction, an inner planet is closest to Earth. At superior conjunction, that planet is farthest; it lies on the opposite of the sun. If an outer planet lies in this latter position, it is simply said to be in conjunction with the sun. (An outer planet at its closest to Earth is in **opposition** to the sun—a relationship of some importance to Copernicus's unlikely debtors, the astrologers, for if one planet enters, say, Scorpio while another happens to be passing through Taurus, they are separated by 180 degrees on the celestial wheel, and fall in contradiction; hence their relationship is inauspicious.) In all of these cases, three heavenly bodies lie in a single line, making the positional calculation of the third a literally straightforward matter to an observer on the first (if he could only see it!). No matter; it's always difficult when you believe in uniform circular motion and lack a telescope . . .

We now know that at superior conjunction Venus is 160 million miles from us; at inferior conjunction, 26 million miles. (In the case of Mercury those distances are respectively 136 and 50 million miles.) Surely such large positional discrepancies should have been suitable for one of Ptolemy's precursor skywatchers to make use of? Copernicus states that the positions of Saturn, Jupiter and Mars can be measured—indeed, can *only* be measured—"when they are in opposition

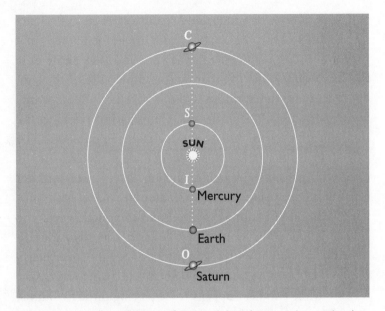

Figure 11 Conjunction and Opposition (Copernican View)
At conjunction, the angular elongation between the sun (or, more generally, any other celestial orb) and the planet in question is zero as seen from Earth.

At inferior conjunction (point I), an inner planet is closest to Earth.

At superior conjunction (point S), that planet is farthest from Earth.

At point C, an outer planet is in conjunction with the sun when it is farthest from Earth.

When it is closest to Earth (point O), it is in opposition to the Sun. Only then does an outer planet retrogress.

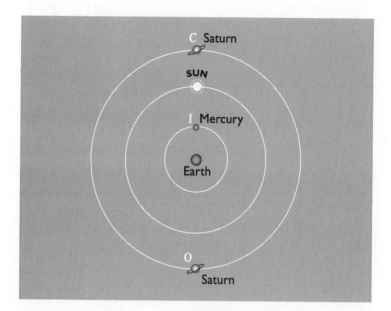

Figure 12 Conjunction and Opposition (If Ptolemy Had Been Correct)

to the sun," and accordingly "lay aside their parallax."* From our terrestrial point of view, we can of course be in line with the two inner planets, but at that point, as Copernicus implied, "they are hidden," obscured by the sunlight, which is powerful enough to cause blindness.

Another perfect circle

Where then is Venus? We have already seen that Ptolemy placed this planet, and Mercury, too, higher than the moon

*A late-twentieth-century guide to celestial navigation for mariners advises us that with the exception of the moon, no parallax correction is required for any of our solar neighbors.

and lower than the sun. Here is his most persuasive reason for doing so in the absence of better certainty: He prefers to use "the sun as a natural dividing line between those planets which can be any angular distance from the sun and those which cannot but always move near it."

By "those which cannot but always move near it" he means the two inner planets, which in Copernicus's words "are not occulted by the approach of the sun, as the higher planets are; and they are not uncovered by its departure. But, coming in front, they mingle with the radiance of the sun and free themselves."

One astrohistorian sums up the Venusian appearances thus: "Venus is drawing in toward the sun, until it is lost in the sunbeams. Then the planet emerges on the other side, not to be seen as an evening star, but as a morning star. In fact, it was plain that in some ways Venus accompanied the sun in its annual movement."

To quantify this accompaniment we now introduce the notion of the **angular elongation** of a planet—namely, the angle along the ecliptic plane between that planet and the sun, as measured from the Earth, in degrees east or west of the sun.

As observed from any church-tower on Earth, Venus's maximum elongation is about forty-five to forty-seven degrees, and Mercury's even less (twenty-eight degrees). The other planets can, as Ptolemy noted, cycle through any and all angles. If we glance at a diagram of the inner solar system as we now perceive it, the reason for this striking difference between the outer and inner planets is visually obvious— heliocentrically obvious, as I should say.

What is Copernicus's word on this matter? "How unconvincing is Ptolemy's argument that the sun must occupy the middle position between those planets which have the full

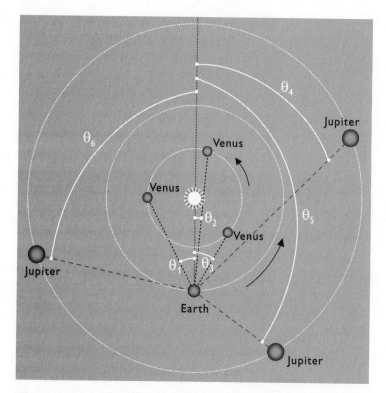

Figure 13 Angular Elongation (Copernican View)

The angle between the celestial orb in question, a terrestrial observer's point of sight and the sun, expressed in degress east or west.

As shown in the Venusian angles θ1, θ2, and θ3, the elongation of an inner planet must always be acute. In the case of Venus, it never surpasses about 47 degrees.

Jovian elongations θ4, θ5, and θ6 show that the elongation of an outer planet can be any value from 0 to 360 degrees.

(For Ptolemy's view of the elongation of an inner planet, note the restricted orientation at the epicyclic diameter in Figure 15. The Copernican interpretation is a wonderful simplification.)

range of angular elongation from the sun and those which do not is clear from the fact that the moon's full range of angular elongation proves its falsity."

This attack, while unsettling, hardly constitutes a *coup de grace.* Thanks to hindsight, Ptolemy's translator puts the case better: "This distinction between the two kinds of planets— those whose elongations from the sun are limited and those whose elongations are not—is accidental in the Ptolemaic system but follows necessarily from the first premises of the Copernican."

It is all too easy for a post-Copernican to interpret these data (correctly) as showing that Venus and Mercury must be closer to the sun than we, and that Mercury must be closer than Venus. Of course, if the old universe were still alive, such an assertion would have made it scream.

Accordingly, within *The Almagest* we find one more of the usual deferent-epicycle-equant systems: A perfect circle revolves not quite around the Earth, carrying with it the center of another revolving circle which traces out the Venusian orbit. Conveniently for Ptolemy (and incidentally for Copernicus), Venus's true orbit turns out to be closer to perfection than that of any other member of our solar system! To be precise, its **orbital eccentricity,** or deviation from circularity, is but 0.007.

All the same, the translator cannot escape the duty of remarking that "within the Ptolemaic setup, it was remarkable and unexplained that the period of the epicycle-bearing circles of Mercury and Venus should be equal to a year and the sun should always be on a line with the center of the epicycle," which was not at all the case for the outer planets. No matter. Can a post-Copernican explain why each planet

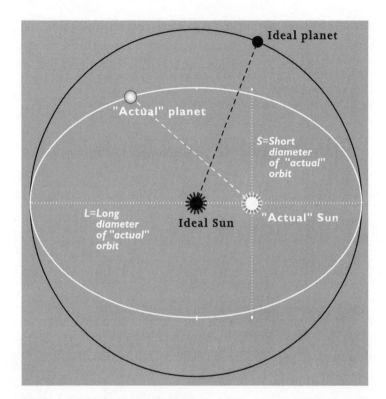

Figure 14 Orbital Eccentricity

L = long diameter of planet's orbit.

S = short diameter.

For the Ptolemaic (and Copernican) ideal of a planet, L = S. In reality, perfect circles are hard to come by. Deviation from the ideal = eccentricity, measured as follows: E = (L− S)/(L + S)

Data for the five planets know to Copernicus:

Least eccentric: Venus at 0.007

Most eccentric: Mercury at 0.206

(Earth is at 0.017)

(excepting Neptune) happens to be almost exactly twice as far from the sun as its inward neighbor? This phenomenon, called Bode's Rule, remains inexplicable. From Ptolemy's viewpoint, therefore, why couldn't the epicyclic peculiarities of Mercury and Venus be no more or less than peculiarities?

"Then what will they say is contained in all this space?"

Kuhn reminds us that the condition that these angles of maximum elongation be forty-five-odd degrees "completely determines the relative sizes of the epicycle and deferent." In *Revolutions* our challenger accordingly concludes that the geometry of Ptolemy's Venus requires an absurdly large domain. "Then what will they say is contained in all this space, which is so great as to take in the Earth, air, aether, moon and Mercury, and which moreover the vast epicycle of Venus would occupy if it revolved around an immobile Earth?"

"With it being so big," Jensen remarks here, "it starts to seem like more than just a mathematical convenience to get the right motion. And indeed Copernicus, though he sticks with circular motion, doesn't have to resort to any epicycles so large, I believe." And so Jensen is sympathetic to this method of attack. Venus's epicycle is *disproportionate* to that of the other known planets. I myself sense something desperately off the mark. Never ceasing to uphold the notion of a finite universe, he demands to know how Venus could enclose so great a space, while never seeing the immensely more disproportionate space which gapes between us and the nearest star in Capricorn. What will any of us say is contained in all this space? What *can* we say?

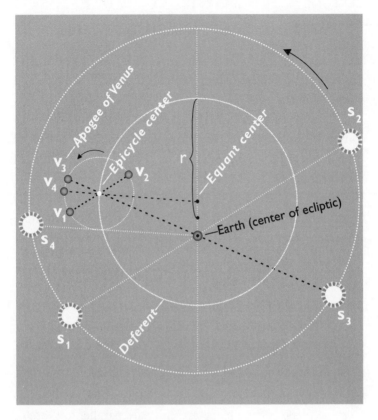

Figure 15 Ptolemy's Venusian Orbit
When Venus is at V_3, it is in conjunction with the mean sun at S_3.

When Venus is at V_1, the sun is at S_1. When Venus is at V_2, the sun is at S_2. In these and all other cases (excepting V_3 and S_3), the line from the epicycle's center to Venus parallels the line from the Earth to the mean sun.

"An easier and more convenient demonstration"

From our thoroughly uncentered point of view, Copernicus's description of Venus's orbit resembles Ptolemy's as much as it differs from it.

"The movement of Venus," says *Revolutions*, "happens to be compounded of two regular movements, either by reason of the epicycle of an eccentric circle, as above, or by any other of the aforesaid modes. This planet however is somewhat different from the others in the order and commensurability of its movements; and, as I opine, there will be an easier and more convenient demonstration by means of the eccentric circle of an eccentric circle."

Circles around circles! This sounds all too familiar.

"More complicated than the Ptolemaic system"

His Venusian orbit is immortalized pearls of ghastly syntax, so let him yield to the astronomer Jacobsen, whose admiration is not entirely unqualified: "Unlike Ptolemy, whose aim was to keep the planes of the planetary epicycles parallel to those of the ecliptic, Copernicus aimed to keep them parallel to the (inclined) planes of their deferents. However, all the lines of **nodes** of the deferents passed (erroneously) through the *mean* annual position of the sun. This device alone, quite apart from introducing errors in distances, gave rise to unmanageable discrepancies in the heliocentric latitudes." Struggling to make everything come out right, Copernicus accordingly invented waverings in *all* extraterrestrial deferents, compounded by periodic swings for some planets, topped off for Mercury and Venus in particular by another kind of tilting called, appropri-

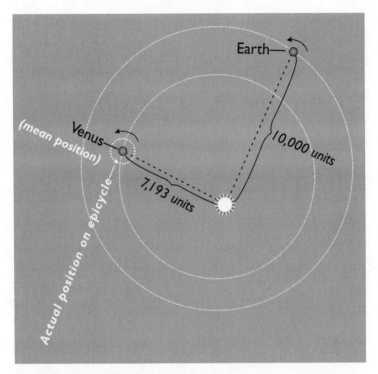

Figure 16 Copernicus's Venusian Orbit
Simplified. Venusian and terrestrial orbital circles are eccentric.

ately enough, **deviation**. Therefore, Jacobsen concludes, "the Copernican system was if anything *more* complicated than the Ptolemaic system, even if its description involved a somewhat smaller number of circles."

"But now the Telescope manifestly shows these horns"

Copernicus has made a botch again, it would seem. But one more matter remains to be told.

Revolutions tells us again where Venus isn't: "As the followers of Plato suppose that all the planets—which are otherwise
dark bodies—shine with light received from the sun, they
think that if the planets were below the sun, they would on
account of their slight distance from the sun be viewed as only
half—or at any rate as only partly—spherical. For the light
which they receive is reflected by them upward . . . toward the
sun, as we see in the case of the new moon or the old."

Yes indeed, Ptolemy and Copernicus both agree with the
Platonists that the moon revolves around the Earth, and that
it derives its light from the sun. Pure logic accordingly asserts
that the lunar phases occur as follows: When the moon is new,
Earth is facing the side which is farthest from the sun. At this
time, "the light which it receives is reflected by it upward."
When the moon is full, Earth faces the side which is facing the
sun. A first-quarter and third-quarter moon each face us at a
90-degree angle to the lunar-solar line. Between the first
quarter and fullness, and again between fullness and the
third quarter, when the moon is gibbous, the angle between
the terrestrial-solar and terrestrial-lunar lines of sight obviously varies between 90 and 180 degrees.

Since, as has been said, Ptolemy finally decided to follow
his forbears in placing the Sphere of Venus closer to us than
the Sphere of the Sun, and since no observation could ever
detect that Venus got more than forty-seven degrees away
from the sun, it became logically impossible for anyone on
Earth to see more than half of Venus's sunlit surface at any
one time. A Ptolemaic Venus must therefore be a crescent
Venus *at most.*

But Copernicus says that Earth moves around the sun, and

so does Venus. Therefore, we ought to be able to see Venus in different phases.

In Copernicus's lifetime, the limits of observation will stifle any proof. But in 1611, Galileo's telescope will bear him out, showing a gibbous phase.

In 1949 a scientist in another discipline defined "the ultimate safeguard of all scientific method. It must meet the test of constant use and re-use, and prove itself sound, not once, but a thousand times. It becomes part of a structure of knowledge and method which must bear a constantly growing burden. Inevitably, a weak unit will crumble and betray itself."

Copernicanism has weighed down Ptolemaic theory and broken it at a weak point. *The Almagest* has finally betrayed itself.

There is a pervasive fable that Copernicanism's founder predicted this outcome. Rosen has written a special essay to puncture it, remarking: "But what about Venus's phases in Copernicus's own system? Of course he never saw them." And at the very end he grimly announces: "Copernicus expressed no opinion about phases of Venus." No matter. Copernicus may or may not have drawn conclusions about the necessary apparent shapes of Venus. All the same, a gibbous Venus was the logical consequence of his system, and another terrible gash in the Ptolemaic universe.

Having observed this gash, however, Galileo continues: "Another and greater difficulty does Venus exhibit; for if revolving around the sun, as Copernicus affirms, it were sometimes above, and at other times below it, receding and approaching with respect to us so much as the diameter of the circle described would be, then at such a time as it is below the sun, and nearest to us, its disk would show little less than forty times bigger than when it is above . . . nevertheless, the difference is almost imperceptible."

And one starts to believe that once more Copernicus has failed, that in yet another crucial respect reality is not as his geometric broodings demand it to be, until several pages later, Galileo informs us: "But now the Telescope manifestly shows these horns" of Venus "to be as terminate and distinct as those of the moon and appear, as it were, parts of a very large circle, near forty times greater . . ."

Modern figures for the variation in Venus's apparent diameter are ten to sixty-four minutes of arc. (Please recall the immense variation, quantified earlier in this chapter, in the distance between Venus and our planet.) Since the area of a circle is pi times the square of the radius, 78.54 to 3216.99 will give the ratio of apparent areas, which works out, I am happy to report, to 1:40.96.

And so once again we find that for all his dreary errors, Copernicus is more true than any astronomer who has come before. "We had a full view at Frauenburg."

Figure 17 Phases of Venus (Drawn to Scale)

Exegesis

Book II

"**N**ow we shall fulfill our promise by proceeding from the whole to the parts," writes Copernicus merrily, and the remainder of *Revolutions* does just that, in thought-explorations which resemble the squat walls and heavy windings of Cracow's streets—and in prose of incomparable dreariness. Kuhn remarks that had *Revolutions* ended with Book I, the Copernican Revolution would not and should not have borne Copernicus's name. The "detailed technical study" of Books II through VI "is Copernicus's real contribution." All the same, I note that Kuhn ends his summary of *Revolutions* at the end of Book I. This essay of mine cannot do much better.

II.1–2: Uncentering definitions

Copernicus begins Book II by reintroducing all the heavenly circles we've already met in *The Almagest*, such as celestial equator and ecliptic. But he de-Ptolemicizes them, inverting them in consequence of his moving, uncentered Earth. His admonition at the end of II.1 is typical:

> But these two circles which have their centers on the surface of the Earth, i.e., the horizon and the meridian, are wholly consequent upon the movement of the Earth and upon our sight at some particular place. For the eye everywhere becomes as it were the center of the sphere of all things which are visible to it on all sides.
>
> Furthermore all the circles assumed on the Earth produce circles in the heavens as their likenesses and images.

He next advises us (and this again derives from *The Almagest*) how to build a simple instrument capable of measuring the sun's shadow. We will need a warp-proof square each of whose sides is about two meters long. Next draw the quadrant of a circle—that is, one-fourth of one—whose center is one corner of the square, and whose radius is one of the square's sides. Subdivide its arc into its 90 standard degrees, and subdivide each degree into 60 minutes; this is why the square needed to be so large. Emplace a nicely lathe-turned "cylindrical pointer" at the center, "and fixed in such a way as to be perpendicular to the surface and to extend out from it a little, say perhaps a finger's width or less." (How precise do these instructions sound to you? "If I could bring my computations to agree with the truth to within ten degrees, I should be as elated as Pythagoras.")

And how shall we employ the new toy? "The next thing to do is to exhibit the line of the meridian on a piece of flooring which lies in the plane of the horizon" and which has been made as absolutely level as possible. The meridian line will be "exhibited" quite conveniently on it as follows: On the piece of flooring, draw a circle whose center is marked by a vertical cylinder. Wait for a sunny day and watch the cylinder's shadow. At sometime in the morning it will touch the circle at one point; in the afternoon it will do so at another. Mark both places on the circle, and bisect the arc between them. The line from the center of the circle through that center point runs north and south—or, in Copernicus's vernacular, it is the meridian.

We can now affix our instrument to this surface with the quadrant's center lying southward along, and at a right angle to, the meridian.

On the summer solstice, mark the noon shadow of the cylinder where it falls upon the quadrant's curved rules. Wait half a year until the winter solstice comes; then repeat the same. The arc between these two points will be 46° 54'.* Since the equinoxes are the points where the ecliptic crosses the celestial equator, the two solstitial points, called the **tropics**, must indicate the ecliptic's maximum distance *away* from the celestial equator. Indeed, these lie directly north along the meridian line, and therefore farther away from each other than any other two shadow-marks. We uncentered Copernicans and post-Copernicans assert that the Earth wobbles off the celestial equator in one direction, crosses it, wobbles past it in the other direction, crosses it again, and returns to its starting point. This motion is symmetrical; therefore the two tropics are equal and may be derived by taking half of 46° 54', which is 23° 27'.

---

*In fact Ptolemy measured a value of between 47° 40' and 47° 45', while Copernicus, as we shall see, detected a lower value, thereby discovering that the obliquity of the ecliptic alters. See below, p. 166.

II.3–14: Tables and transformations

Copernicus constructs his spherical triangles and determines angles and relative distances. Now comes a table of declinations of the degrees of the ecliptic and a table of right ascensions—essentially, celestial latitude and longitude: "The angle is right where the meridian circle by definition cuts the equator described through its poles. Now the arc of the meridian circle, or any arc of a circle passing through the" celestial "poles and intercepted in this way is called the declination of a segment of the ecliptic; and the corresponding arc on the equator is called the right ascension."

It is astonishing how much Copernicus and his predecessors have been able to accomplish with such limited means. For instance, *Revolutions* informs us that "for any given altitude of the sun the length of the shadow is derivable and vice versa." "The differences between right and oblique ascensions are the same as the differences between the equinox and a different day." "When some degree of the ecliptic is given, the rising of which is measured from the equinox, the degree which is in the middle of the heavens is also given."

Book II is a virtuoso repertoire of such transformations, whose language becomes so abstract that it sometimes resembles poetry: "Wherefore *EN* is half the chord subtending twice the arc of the horizon which is the difference between sunrise on the parallel and the equinoctial sunrise."

Again he sums up the practical utility of his operations: "Hence the risings and settings can be easily understood."

He endeavors to tame the apparent wriggles of the ecliptic, noting that "if we take the right ascension corresponding to the known degree of the sun and if for every equal hour measured from noon we add 15 degrees 'times' to it . . . the sum of the right ascensions will show the degree of the ecliptic in the middle of the heavens at the proposed hour."

In his murky fashion he endeavors to explain how to construct an astrolabe, whose revolvable, degree-marked circles comprise a sort of microcosm of the Ptolemaic cosmos. There is no space here to detail its construction and use. Suffice it to say that one begins by setting the ecliptic circle to the sun's known position at the moment of observation, then takes a reading on the moon to determine its longitude, "for without the moon there is no way of discovering the positions of the stars, as the moon alone among all is the partaker of both day and night." Then one sights upon a given star to obtain its celestial longitude and latitude.

For this device, and for myriad observations taken with it, he gives thanks to Ptolemy, whom he calls "that most outstanding of mathematicians," but notes that his predecessor located the stars in relation to the spring equinox, which, since the latter alters with precession, should rather in his (Copernicus's) opinion be described in relation to the Sphere of Fixed Stars. "Actually," Eric Jensen comments here, "astronomers still use a system in which the zero point of right ascension is set by the spring equinox. Since that *does* move with precession, as noted, so too do the coordinates of the stars. Thus, when I quote the coordinates [right ascension and declination] of a star in order to communicate its position to other astronomers, I also have to give the year for which I'm assuming the position of the vernal equinox. There are standard equinoxes that astronomers use, and the transformations between them are straightforward now that the rate of precession is known with high precision." In Copernicus's epoch, of course, that was not quite the case, and I for one cannot blame him for preferring a motionless, eternal system. Alas! Such fancies went out with centeredness.

Enough. He ends Book II with a longish catalogue of the longitudes and latitudes of the constellations' stars.

What We Believed

Scriptures

In this deficient sublunary realm of ours, it is possible to disprove a point, but impossible to prove it eternally and with perfect certainty. All we need do to invalidate Y is find a flaw in the logic which connects it to its primary cause X. But if Y, X, W and all their kindred deductions and syllogisms all the way back to Postulate A exist within a self-contained, self-sufficient sphere of perfect consistency, further evaluation may become irrelevant to the faithful.

The parable of the lodestone

A postulate is an assumption we begin with. Therefore, it lies beyond proof, although one can't ever exclude the chance that some fancy logic-chain will allow it to be supported or invalidated by a different postulate. (Leibniz: "One should not omit any necessary premise, and all premises should have been either previously demonstrated or at least assumed as hypotheses, in which case the conclusion is also hypothetical.") Our postulate may be upheld by every datum we have ever gathered, but God makes no guarantee that our discoveries tomorrow will continue to uphold it. "The limits of observation" deserved their own chapter and got one; for one main theme of the Copernican Revolution is the laborious enlargement of human vision. But only sentimentalists can pretend that our story consists of the straightforward substitution of Truth for Error. For instance, Newton postulated, quite reasonably given his epoch's level of scientific knowledge and quality of measuring instruments, that gravity and other forces act instantly and universally, in the same way at the same time. But almost two centuries later, Maxwell showed experimentally that a lodestone does not draw iron towards it at once; the interval between the introduction of the magnet and the motion of the iron can be measured. Newton's postulate was shattered. Hurrah; Truth has replaced Error! But how long will Maxwell's own "hypothetical conclusion" survive unmodified? Who can say?

God can, if one believes in God.

Exempt from re-examination

Most of us cling to any number of postulates which we've rendered exempt from re-examination—meaning not only what my old calculus textbook describes as "a formal statement that is assumed to be true without proof," but also "founded on premises exempt from definition." Here are some of them: *I'll love you forever. My country is always right. All men are created equal. My sensory perceptions of physical phenomena are as accurate as they need to be. Science has the right to inquire unrestrictedly into everything.*

God exists. We cocoon them in consistency. I may succeed in demolishing one of these to my own satisfaction without even scratching it in your estimation. In spite of Darwin, Creationism continues to survive quite nicely; in spite of Nazi atrocities, a few Germans go on believing that *my country is always right.* They may deny the Holocaust, or they may justify it by means of some other axiom immune to re-examination, such as *the Jews are our misfortune.*

This book deals with a moment in the history of science when reasoning was *by its own logic* subordinated to religious faith. A long moment it was, but still a moment; for over time, the Church's stance on cosmology and astronomy altered; meanwhile, learned opinions predictably differed. In 1215 we find the Fourth Lateran Council attacking Aristotle's doctrines. A century later, Nicole Oresme criticized many of Aristotle's arguments concerning motion and other matters. But the long and short of it is that once Christianity had established itself securely enough for pagan writings to pose no threat, it began to find the Aristotelian universe convenient. The correspondence grew almost perfect. Even the point-like nature of our Earth, mentioned by Ptolemy and Copernicus, finds its Scriptural analogue. Hence Augustine: "Thou created Heaven and Earth; things of two sorts: one near Thee, the other near to nothing."

All the same, the Church's opinion regarding the scientific logic which had brought the Aristotelian universe into being in the first place remained unpredictable. Kuhn asserts that "before the tenth century and again after the sixteenth," namely, Copernicus's century, "the Church's influence was, on balance, antiscientific."

Ptolemy and Aristotle divide the sciences into three: the

theological, through which we can hope to apprehend "the first cause of the first movement of the universe," the mathematical, which resides "in absolutely all beings mortal and immortal," and the physical, which examines "the white, the hot, the sweet, the soft, and such things." The theological "is in no way phenomenal or attainable" (read "exempt from re-examination"), the physical, in contradistinction to our modern uncentered ideas of it, "is unstable and obscure, so that philosophers could never hope to agree," which leaves only the mathematical able to "give its practitioners certain and trustworthy knowledge with demonstrations."

Upon this triad, the Church imposes itself. Some years before *Revolutions*, in a dialogue entitled *De Deo abscondito*, Nicolas of Cusa has a Christian explain to a pagan:

> We worship truth itself, absolute and unadulterated, eternal and ineffable, but you, erring as you do, worship not truth itself, absolute as it is in itself, but as it exists in its works; you worship not absolute unity but unity in number and manyness,

which sounds not terribly distinguishable from the hot, the sweet, the soft, and such things unstable and obscure.

Not only did dissent from *God exists* become unspeakable (and dangerous) but there was then (as there is now) no *scientific* reason not to believe it.

Then as now, I said. The *Twentieth Century Encyclopedia of Catholicism* informs us that "however the science of cosmology should develop in the future, a person's belief in the truth or falsity of the Genesis account of an initial creation, *properly understood*, will be unaffected." And the *Encyclopedia* is correct.

But then was not quite the same as now, for the *Encyclopedia* feels compelled to add that "properly understood."

Then as now, *God exists* was one of those postulates which had become exempt from re-examination. But there was no "properly understood." *The Scriptures had become literally true.*

The status of the sun when Lot came to Zo'ar

The most literalist reading of Scripture could be made compatible with scientific discoveries as most of us now understand them. There appear to be three types of astronomical (or cosmological) assertions in the Bible. The first is the most common. It is a factual description of heavenly phenomena or cosmology, told in the prescientific language of the time: "The sun had risen on the Earth when Lot came to Zo'ar." This need not be interpreted to mean that the sun rotates around the Earth; any astonomer can refer to sunrise without being considered a fool; "sunrise," like any idea, is shorthand. As Copernicus himself remarks on this point: "We are speaking in the usual manner of speech which can be recognized by all." And *Revolutions* itself is rife with such conveniences as: "In the case of the horizon . . . the parts of the world," meaning universe, "have their risings and settings on it." Science owns no *a priori* interest in denying that Lot could have come to Zo'ar after sunrise.

In the second place, one encounters the purely visionary. In the Book of Revelations, the seven stars in Christ's right hand exist beyond science, history or anything which we have ever known. A Christian, literalist or not, remains at logical liberty to believe that the laws of astronomy and physics which have been so reliable up until now will come to an end, *at* the End. Science cannot prove that they will not. Science can only *predict*, based on prior observation, which Scripture rules out as irrelevant, that it will not. This is what the *Encyclopedia of Catholicism* is getting at when it opines that "one cannot say anything about an initial creation of theological interest by reference to the telescope, neither can one use astronomy to provide a revised version of Genesis . . ." What if Earth and sun do someday reverse roles at Zo'ar?

Finally, there are moments when the Bible becomes specifically allegorical, as when Joseph dreams that the sun, moon and eleven stars have made obeisance to him, at which his father scolds: "Shall I and your mother and your brothers indeed come to bow ourselves to the ground before you?" How can we exclude from consideration the possibility that the rising of the sun at Zo'ar might comprise just such a parable?

Unfortunately, as convenient as these three categories seem to me, my own inferences as to which passages should be placed in which pigeonhole may well differ from another reader's. When the Book of Job informs us that God set the Earth in a foundation and laid its cornerstone "when the morning stars sang together," my inclination is to see this as part of God's extended question to Job: "Where were you when all this happened? How do you dare to think you know anything?" In which case, why should I think I know anything about the creation of the Earth? Surely God must be employing allegory here—a beautiful one—so that we can begin to understand what little (and how little) we *can* understand. This is how I proceed when I try to explain some practical, mechanical or scientific principle to a small child: The car needs gasoline because it is thirsty now, and gasoline is all that it can drink. If it doesn't get gas, it will soon be tired and then it won't be able to take us home. How can such explanations *not* partake of Aristotelian volitionalism? The car is thirsty and the morning stars sang together. But that is only my isolated, hence impoverished explication; someone else might conclude that the stars literally sang.

"The sun had risen on the Earth when Lot came to Zo'ar." To me, that's vernacular. To Martin Luther, it's literal fact.

A scientific text asserts its own literal truth. When Copernicus writes that the Earth circles around a point in space which is very near the sun, he asks us to accept this exactly, no more, no less. Commenting on Osiander's preface, an astronomer sums up Copernicus: "In reading his words throughout *Revolutions*, one cannot doubt that he believed in the possibility, even the probability, of his system being a physical reality," a literal reality which Osiander mendaciously denied.

The inductive results of these considerations are:

POSTULATE: SCRIPTURAL TRUTH MAY BE EITHER LITERAL OR METAPHORICAL.

POSTULATE: SCIENTIFIC TRUTH CAN BE ONLY LITERAL.

COROLLARY: ACCORDINGLY, WHEN SCIENTIFIC REASONING IS BASED ON SCRIPTURAL EVIDENCE, IT MUST INSIST, RIGHTLY OR WRONGLY, ON THAT EVIDENCE'S LITERAL TRUTH.

This last corollary is responsible for much of the tragedy which befell Copernicanism.

"Aided by spiritual insight"

Because the Church played an evil part in the Copernican episode, we must take care to avoid assuming that Scriptural astronomy played an equally foolish, baneful part.

One of Copernicus's contemporaries, a Milanese scholar who wrote treatises on everything from poisons to Ptolemy, from ether to Nero, from dreams to morals, from the Virgin to the urine, defines an "aspect of knowledge called *proof* because it is derived from the effect based on the cause." You and I would agree in our own uncentered way. "However," continues the Milanese, "in this field of understanding I have less rarely arrived at comprehension by a skillful treatment than I have been aided on many occasions by spiritual insight."

What is axiomatic truth, if not a beautiful mental pattern? The nurturing power of Scriptural astronomy may be indicated by the fact that Kepler, even while upholding the now anticlerical heliocentric theory of Copernicus, called on the following inspiration in considering the relative positions of the sun, the stars and the space between them: the Father, Son and Holy Ghost. "I shall pursue this analogy in my further cosmological work."

As for Copernicus himself, he justifies his heliocentric bravery on the grounds, irrelevant enough in our godless cosmos, of *appropriateness*: "In the center of all rests the sun. For who would place this lamp of a very beautiful temple in another or better place than this wherefrom it can illuminate everything at the same time?" And he goes on to exclaim: "How exceedingly fine is the godlike work of the Best and Greatest Artist!" Why not consider *Revolutions* itself a work of Scriptural astronomy?

Twenty-four centuries since Creation

All the same, I repeat that *when scientific reasoning is based on Scriptural evidence, it must insist, rightly or wrongly, on that evidence's literal truth.* That is why in Copernicus's epoch, it was far from uncommon for a scholar to write something like this: "I have summed up the history of 2,454 years from the beginning of the world" until now (ca. 1570) "more briefly than the magnitude of the material deserves." Most scholars of my own time estimate the Earth's history as running some 4.5 billion years, based on scientific observation and theories not available to the just quoted pious Lutheran Martin Chemnitz. But if we could have introduced Chemnitz to the fossil record and all the rest of it, would he have reacted any differently from the worthy savants who refused to look through Galileo's telescope? As he warns us, "Our reason exalts itself against the knowledge of God."

Meanwhile, Scriptural literalism, which derives (one hopes) only from the noblest respect for God and the most loving humility of reason, makes its own misinterpretations and likewise exalts itself against the knowledge of God. "The sun had risen on the Earth when Lot came to Zo'ar." Who are we to insist that this means that the sun revolves around the Earth? Who are we to be certain that we understand the word of God?*

---

*The Scriptures are literally true. Hence that passionate rebel Galileo defends his Copernican views as follows: "Our opinion is that the Scriptures accord perfectly with demonstrated physical truth. But let those theologians who are not astronomers guard against rendering the Scriptures false by trying to interpret against them propositions which may be true and might be proved so."

Axioms of Scriptural astronomy

It has been remarked that the survival of rigid Aristotelianism for close to two millennia was the result "not so much in the astronomical doctrine as such, but in the steady absorption of the doctrine with the current religious views." In fact, the Scriptures and Aristotelianism reinforced one another to an impressive extent.

When God created the Earth, He gave us explicit dominion over it, and instructed us to "be fruitful and multiply, fill the Earth and subdue it." He certainly did not give us dominion over Heaven—quite the contrary. All the same, the celestial bodies were placed there for our benefit and guidance, for God brought them into being by commanding: "Let there be lights in the firmament of the heavens to separate the day from the night; and let them be for signs and for seasons"—for instance, do not sell grain during the new moon—"and for days and for years, and let them be lights in the firmament of the heavens to give light upon the Earth." It seems that we are neither to approach them (look what happened to the builders of the Tower of Babel), nor by any means to worship them; they are simply the servants of God: "Thou hast made the moon to mark the seasons; the sun knows its time for setting." Such assertions reinforce

AXIOM 1: WE INHABIT A CENTRAL, MOTIONLESS EARTH ABOUT WHICH ANCILLARY HEAVENLY BODIES ROTATE IN PREDETERMINED PLACES.

In the Book of Job, some of God's attributes receive mention: He is the One "who shakes the Earth out of its place, and its pillars tremble, who commands the sun, and it does not rise, who seals up the stars." I take it that the sun does rise unless commanded by God—one more indication of its subservience to Him—and that the Earth has a place to be shaken out of. Likewise, when Christ warns His Disciples of the Second Coming—"the sun will be darkened, and the moon will not give its light, and the stars will fall from heaven"—the heaven which the stars fall from sounds like another ordained place. Such passages hint at a cosmography which may well be as specific as geography but which is evidently not ours to know.

(Evidently the Church does know, for Father Tolosani, whom we've quoted attacking *Revolutions* for violating Aristotle's rules of motion, informs us that "Copernicus would have spoken correctly, had he agreed with the theologians that above the First Moveable," which turns just above the Sphere of Fixed Stars, "the highest sphere is immovable, the sphere called by the theologians the Empyrean Heaven.")

COROLLARY 1: ONE OF THESE ROTATING BODIES IS THE SUN.

Among the Scriptural "proofs" for this, the beautiful passage in Ecclesiastes stands out: "The sun also rises and the sun goes down, and hastens to the place where it rises." Again, Isaiah prophesizes that by way of sending a sign, God will turn back the shadow on the sundial by ten degrees. "So the sun turned back on the dial the ten steps by which it had declined." Scripture further informs us that God "makes His sun to rise on the evil and on the good and sends rain on the just and on the unjust." Rain literally does fall, and the coupling of the falling rain with the rising sun makes the latter more literal, too.

Finally, here is the famous passage in the Book of Joshua which Martin Luther called down on Copernicus's head:

Then Joshua spoke to the Lord . . . and he said in the sight of Israel, "Sun, stand thou still at Gibeon, and thou Moon, in the valley of Ai'jalon." And the sun stood still, and the moon stayed, until the nation took vengeance on their enemies . . . The sun stayed in the midst of heaven, and did not hasten to go down for about a whole day.

COROLLARY 2: THE PREDETERMINED PLACES OF THE HEAVENLY BODIES ARE SPHERES OR SHELLS.

About two hundred years after the birth of Christ, a certain Irenaeus writes in his *Proof of the Apostolic Teaching* that "the Earth is encompassed by seven heavens, in which dwell Powers and Archangels": Wisdom, Understanding, Counsel and so on down to "this firmament of ours, full of the fear of this Spirit, Who lights up the heavens." Not long before A.D. 500, the Pseudo-Dionysius the Aeropagite postulates nine ranks for his *Celestial Hierarchy.* Anyone who's heard of the eight Ptolemaic spheres might well find this schema comfortingly familiar. Needless to say, in the higher ranks, the superlunary ones I should say, one finds "transcendence over every earthly defect" since they are so close to God; how could any pious soul believe that perfection decreases as we approach Him? Since we are so imperfect, then surely the heavenly bodies must be more perfect than we. Ptolemy and Aristotle assert exactly this.

AXIOM 2: THE HEAVENLY BODIES TURN OF THEIR OWN WILL.

The motive powers and consciousnesses of the heavenly bodies never get definitively stated in the Bible. However, we do know from Job that the stars sing, literally or not, which supports the Aristotelian doctrine of the volitional motion of natural bodies. Moreover, if the stars lacked volition we might be forced to abandon astrology—in which case how could our royal counselors plot the most favorable instant for launching wars and suchlike courses of statecraft?

For his part, Ptolemy expresses a viewpoint of crucial relevance to Scriptural astronomy:

> That special mathematical theory would most readily prepare the way for the theological, since it alone could take good aim at that unchangeable and separate act, so close to that act are the properties having to do with translations and arrangements of movements, belonging to those heavenly beings which are sensible and both moving and moved, but eternal and impossible.

Saint Augustine uses the fact that oil always rises to the surface of a volume of water as illustration of the Aristotelian principle that "the body by its own weight strives towards its own place," which in turn elucidates the Scriptural principle that "our rest is our place. Our love lifts us up thither, and Thy good spirit lifts up our lowliness from the gates of death."

Axiom 3: The heavenly bodies are more perfect than Earth since they are eternal and close to God.

As 1 Corinthians informs us: "Not all flesh is alike ... There are celestial bodies and there are terrestrial bodies; but the glory of the celestial is one, and the glory of the terrestrial is another." (No matter that Corinthians then goes on to distinguish the sun from the moon, and stars from each other.)

This helps "prove" the Aristotelian notion that change and corruption are sublunary, and the stars, planets and moon are comprised of some eternal fifth element, perhaps ether.

COROLLARY (THE PERFECTION RULE): THE PATHS OF THE HEAVENLY BODIES FOLLOW PERFECT GEOMETRICAL PATHS, SINCE THEY ARE ETERNAL AND CLOSE TO GOD. HENCE ORBITS ARE CIRCULAR.

Ptolemy has already weighed in literally on the side of the angels in this regard, informing us that "we believe it is the necessary purpose and aim of the mathematician to show forth all the appearances of the heavens as products of regular and circular motions." He gets reinforced and vindicated by the Wisdom of Solomon (11.21): *Thou hast ordered all things by measure and number and weight.*

AXIOM 5: GOD IS LITERALLY PRESENT.

If so, then where is He?

The Christian God is said to have originated all things, as opposed to so many pagan gods who arose out of water or from Earth itself. Accordingly, He is greater than all, above all, before and after all. If anywhere, it is in the heavens that we should look for Him. Presumably His throne stands in the Empyrean Heaven beyond the outermost sphere of the cosmos—another reason not to argue against the Sphere of Fixed Stars.

One historian describes the pre-Christian universe as "a vast republic of gods, men, animals, plants and things realizing their respective natures and coming to rest in their appointed places, as the eternal forms embody themselves perpetually in new matter according to the heartbeat of life and death." Aside from the compulsory substitution of God for gods, the early Scripturalist universe possessed an equally vitalist character. And what could be more Aristotelian than that?

The leaden square

The Scriptures are literally true. Catholics and Protestants agree on that—but Protestants go farther, while Catholics insist that Scripture is not only incomplete, but (so a Lutheran who demands to differ paraphrases them) "in those things which it does contain it is obscure and ambiguous, like a waxen nose or a leaden square . . ."

In other words, not only are the Scriptures exempt from re-examination, but so is contemporary authority's self-referential interpretation of them.

"Indeed we declare, announce, and define that it is altogether necessary to salvation for every human creature to be subject to the Roman pontiff." These words were composed in 1302, long before Luther, by Pope Boniface VIII. In Copernicus's time, when the Reformation which Luther spearheads has wounded the Roman-led unity of Christendom as much as *Revolutions* will the Scripturalist universe, the leaden square's ambiguity must in self-defense grow ever more stringently denied.

The first capital punishment for heresy is said to have been suffered by the Gnostic thinker Priscillian. How many others since then have perished for sketching out the leaden square in their own dimensions?

Half a century before Copernicus's birth, a Hussite sectarian named Jerome of Prague, laden down with fetters, faces his examiners and utters what now would be considered one of the most fundamental assertions of the scientific credo: "Prove that what I advanced were errors, and I will abjure them in all humility, and most sincerely."

The reply: "To the flames with him! To the flames!"

To the flames he will go, after a year of endungeonment under atrocious conditions. On the eve of his execution, he dares to argue against the Cardinal of Florence that Holy Writ itself is a better guide than the interpretations of the Fathers. The Cardinal turns away in fury. Then they burn Jerome of Prague alive.

Jerome was not, as it happens, a scientist, much less an astronomer. His discourse was more public than Copernicus's, and therefore more threatening. Had he acquiesced to the interpretations of the Fathers even while scribbling heliocentric nonsense, they might have left him alive.

Two years after the publication of Copernicus's *Revolutions of the Heavenly Spheres*, the Council of Trent goes into session.

In 1894 the Bishop of Rottenburg informs us: "That the origin of councils is derived from the Apostolic Synod held in Jerusalem about the year 52 is undoubted; but theologians are not agreed as to whether they were instituted by divine or by human authority."

There are eight kinds of councils, of which the universal or ecumenical are the highest. Bishops and their dignitaries come from all over the world to attend them, as summoned by the Pope or his legate. As might be imagined, it takes a crisis to convoke an ecumenical council: a dangerous heresy, a clash between opposing Popes, consideration of some fundamental reform. Whatever decisions the ecumenicals reach will have force of law for all Catholics, and be considered infallible. The first ecumenical council was that of Nicaea, in A.D. 325; the Council of Trent was either the sixteenth, if we count only uncontested councils, or the nineteenth.

The Council's fourth session, that of 8 February 1546, addresses a question not without relevance to Scriptural astronomy: *Should reaffirmation of the canonical nature of all books of the Bible be given with or without re-examination?* It turns out to be a divisive discussion; but in the end, needless to say, reaffirmation carries the day without an excess of re-examination.

Only the Vulgate, which is St. Jerome's translation, will now be authorized, since there are too many other versions presently in use.

The Cardinal of Jaën proposes that only Doctors and Clerics be allowed to comment on Scriptures, but he is overruled since the Scriptures are for all. I feel happy about this.

Now, what about traditions? The Bishop of Chioggia proposes that Church traditions be considered merely laws, not revelations, but excites an uproar of anger. "The license of interpretation being the great evil of the day," sums up one yellowed history, ". . . it was recommended that it should be forbidden to interpret the Scripture contrary to the declared sentiment of the Church, and to the unanimous consent of the Fathers . . ."

Or, as the Council more blandly resolves, the Synod "venerates with an equal affection of piety, or reverence, all the books both of the Old and the New Testament . . . as also the said traditions, as well those appertaining to faith as to morals, as having been dictated, either by Christ's own word of mouth, or by the Holy Ghost, and preserved in the Catholic Church by a continuous succession."

In anticipation of the next Jerome of Prague, the standard legal penalties for disobedience are enjoined.

"The Sun did run much more than 7,000 miles"

From my own perspective these proceedings seem monstrously unfair. I believe in freedom of thought and self-expression. The dignitaries at the Council of Trent did not. But remember this in their favor: They were faithful to their postulates not only out of loyalty but also because for them it was their science; it was logically true. We would be omitting a crucial part of the story if we devalued these mens' ability to manipulate quantities and qualities just as rationally as we.

Somewhere around the year 725 A.D., in a clammy cell in Northumberland, the Venerable Bede mentions the plausible opinion of the faithful that because in Genesis God divides day and night into two equal parts, "we should believe that the beginning of the world took place specifically at the equinox." Bede differs; he points out that light was created three days prior to the creation of the luminaries without which there could be no equinox. Let's call Bede a scientist. He takes his observations (studies Scripture), interprets what he perceives, and deduces accordingly.

Nine centuries later, Cardinal Roberto Bellarmino shows himself equally capable of calculation and deduction:

> I myself being once desirous to know in what space of time the Sun set at sea, at the beginning thereof I began to recite the Psalm *Miserere*, and had scarce read it twice over before the Sun was wholly set. It must needs be, therefore, that the

Sun in that short time did run much more than the space
of 7,000 miles. Who would believe this unless certain reason did demonstrate it?

Oh, certain reason! How utterly reasonable this Bellarmino
is! He reasons out his leaden squares. "As you [are] aware," he
threatens Galileo's go-between Foscarini some seventy-two
years after Copernicus's death, "the Council of Trent forbids
the interpretation of the Scriptures in any way contrary to the
common opinion of the holy Fathers . . ." And what might
those be? The Cardinal will now tell us: "All agree in interpreting [Scriptures] literally as teaching that the Sun is in the heavens and revolves around the Earth with immense speed . . .
Consider, then, in your prudence, whether the Church can
tolerate that the Scriptures should be interpreted in a manner
contrary to that of the holy Fathers and of all modern com-
mentators, both Latin and Greek."

This contrary interpretation, unfortunately, was precisely
what Copernicus had assembled in *Revolutions*.

Exegesis

Book III

Giving due credit to Hipparchus, who'd discovered that the year which is defined as the time which it takes a given star to complete an annual revolution exceeds the year which is defined as a completion of the annual cycle of equinoxes and solstices, Copernicus observes that the fixed stars are fixed only in relation to each other, so far eastward have they already moved that the twelve positions of the ecliptic no longer correspond to the twelve signs of the Zodiac for which they were named. There's precession for you! Doesn't that call astrology itself into question? Poor Tiberius, who cast horoscopes to see into enemy plots! And let's proffer sympathy to the Mesopotamians who calculated from the position of the Libra the Balance, not from the quantity of wheat produced that season, what wheat prices ought to be!

"Moreover," pursues Copernicus, "an irregular movement has been found." Could the Sphere of Fixed Stars be advancing in enigmatic little jerks? "The head of the constellation of

Aries has become more than three times 8° distant from the spring equinox."

"For the sake of a cause for these facts," he remarks in one of his infrequent attempts at drollery, "some have thought up a ninth sphere and others a tenth: they thought these facts could be explained through those spheres; but they were unable to produce what they had promised. Already an eleventh sphere has begun to see the light of day . . ."

No doubt he himself would yield to compulsion, and add a sphere or two if he absolutely had to. But any so-called "irregular movement" will never win the victory over Nicolaus Copernicus, defender of the faith!

Book III of *Revolutions* is noteworthy for its ingenious attempts to rationalize the apparently nonuniform motions of precession and of the ecliptic's obliquity; it also compares various schemes to compute our Earthly days and years, together with its own conclusions, which obviously aim at the improvement of the calendar: As a result of precession compounded by the "irregular movement," the year as calculated from equinox to equinox must vary. He discusses the observations of Ptolemy and others, including himself, in hopes of finding how long the year actually is; and logically determines that "the equality of the solar year is more correctly measured from the Sphere of the Fixed Stars, as Thebites ben Chora was the first to find." To its length, as computed by Thebites, namely 365 days, 6 hours, 9 minutes, 12 seconds, Copernicus adds 28 seconds, so he is slightly farther than Thebites from our computation of 9.5 seconds, but still impressively accurate.

Most of all, however, Book III claims our notice for the peculiar mixture of old and new it contains: an uncentered Earth, but a nearly centered sun in a finite universe of perfect circles.

III.1–3: Spica's variables

One exasperating example of the latter—and a main part of my reason for haling you through these summaries as quickly as possible—has to do with the way Copernicus tries to get across his point. Nowadays we are used to graphic displays of information. Copernicus seeks to diagram his way through the universe, but his diagrams have been labeled merely with letters, the keys being buried out of order in paragraphs, with abstruse subordinate clauses sprinkled on them like dirt, so that to understand a schematization we sometimes have to struggle through the adjoining paragraph half a dozen times. Perhaps that is why Copernicus receives the following accolade from one of his countrymen: "He knew how to explain the hidden causes of phenomena on principles worthy of admiration."

When he omits to diagram, the case often becomes worse.

Consider his discussion of the precession of Spica. Trying to be helpful, he relates how this blue-white star, which has been classed by astrologers as "venereal and mercurial" and which lies on the celestial equator in the constellation of Virgo, has altered its position over the centuries since the death of Alexander the Great. His figures are a maddening cornucopia of apples and oranges. Sometimes he refers to changes in Spica's angular elongation; other times he gives us Spica's longitude and latitude. Never mind. "Wherefore in the year of Our Lord 1525, in the year after leap-year by the Roman calendar and 1849 Egyptian years after the death of Alexander, we," meaning of course Copernicus, "were taking observations of the often mentioned Spica, at Frauenberg, in Prussia." Its declination had changed by four arc-minutes since our protagonist's previous observation a decade before; its distance from Virgo had altered by seven arc-minutes in the same period. This seems almost insignificant ("if I could bring my computations to agree with the truth to within ten degrees, I should be as elated as Pythagoras"); so let's consider a longer interval: When Ptolemy observed Spica, the star's declination was a mere half a degree as compared to the 1525 declination of 8° 40'.

Now that we've been apprised of all this, how do we calculate Spica's rate of precession? My common sense proposes the following: Subtract Ptolemy's declination value from Copernicus's; divide the difference by 1423, which is the number of years between the two observations, and obtain 0.0057. That's about half a degree per century. Did I comprehend Copernicus? Evidently not, since *he* concludes that the equinoxes and solstices moved eastward about one degree per century in Ptolemy's time, and one degree per seventy-one years thereafter.

Oh, *now* I'm reminded by context that one is supposed to use the tables in Book II to convert declination to ecliptic longitude. "For in the 266 years between Hipparchus and Ptolemy, the longitude of Basilicus in Leo from the summer solstice moved $2\frac{2}{3}$ degrees, so that here too, by taking the time into comparison, there is found a precession of 1 degree per 100 years."

As usual, Ptolemy was clearer.

III.3–4: The lost ellipse

The seemingly irregular alteration of the equinoxes Copernicus explains through "two reciprocal movements belonging

wholly to the poles, like hanging balances," one of them altering the polar tilt, the other acting crosswise. "Now we call
these movements '**librations**,' or 'swinging movements,'
because like hanging bodies swinging over the same course
between two limits, they become faster in the middle and very
slow at the extremes." One of these he calls the anomaly of
declination, the other the anomaly of the equinoxes. "And so
these two librations competing with one another make the
poles of the Earth . . . describe certain lines similar to a twisted
garland."

Faster in the middle! But wouldn't that imply the ultimate
anathema, non-uniform motion? —Never fear! Copernicus
will not betray the faith. "We have to show that when the twin
movements of circles GHD and CFE compete with one
another, the movable point H proceeds back and forth along
the same straight line AB by a reciprocal motion."

In the first diagram he draws, it becomes clear that while
these movements are not carried around on other larger
movements, as epicycles are, all the same, they bear analogy to
epicycles in that they are regular circular resolutions to a
problem of apparent irregularity.

In the second diagram, a circle turns clockwise, its center
lying on the diameter of the innermost of two nested counterclockwise circles. The clockwise circle corresponds to the
anomaly of the equinoxes, which is twice as rapid as the
anomaly of declination. It all reads very plausibly.

Koestler notes that the manuscript version of this section
(III.4) contained the following passage: "If the two circles have
different diameters, other conditions remaining unchanged,
then the resulting movement will not be a straight line but . . .
what mathematicians call an ellipse."

(Actually, points out Koestler, it would be "a cycloid resembling an ellipse.")

"The odd fact," continues this commentator, "is that Copernicus had hit on the ellipse which is the form of all planetary orbits—had arrived at it for the wrong reasons and by faulty deduction—and having done so, promptly dropped it; the passage is crossed out . . ."

III.5–26: Eccentrics, epicycles and an uncentered Earth

The obliquity of the ecliptic has decreased since its measurement was first taken by Ptolemy's predecessors. He finds that its complete cycle lasts 3434 years, and the precessional period of the equinoxes comprises exactly half that, or 1717 years.

Four and a half centuries later, Jacobsen will point out that obliquity is not accurate as *Revolutions* defines it. No matter. Copernicus soldiers on. He tabulates for us precession in longitude over intervals both of days and of Egyptian years; and does the same for the anomaly of the equinoxes.

"Let our problem be to find the true position of the spring equinox together with the obliquity of the ecliptic for the 16th day before the Kalends of May in the year of Our Lord 1525, and how great the angular distance of Spica in Virgo from the same equinox is." If this is the sort of problem you crave to solve, why then, Copernicus is your man.

He calculates and tabulates the regular and mean rotations of the center of the Earth. The fact that these are not precisely centered on the sun "can be understood two ways, either through an eccentric circle, *i.e.,* one whose center is not the center of the sun, or through an epicycle on a homocentric circle." And he draws a diagram of both, while admitting that determining which one best represents the facts is tricky. Fortunately, either case will uphold Ptolemy's dictum that for the Earth, as for the sun, moon and five known planets, "all their apparent irregularities" are "produced by means of regular and circular motions (for these are proper to the nature of divine things which are strangers to disparities and disorders)."

In Copernicus's favor—and Ptolemy's, and Aristotle's—we should remind ourselves that our solar system's planets do not deviate from perfect circularity all that much. For instance, the gas giants such as Jupiter and Saturn rarely surpass orbital eccentricities of 0.1. The worst offender is Pluto (0.25), which in Copernicus's universe thankfully declines to exist. Unfortunately, Mercury and Mars (0.21 and 0.09) also disobey the rules. No matter; an epicyclic adjustment or a reorientation of the deferent can save even *their* appearances.

Pass on to less awkward matters: Thanks to Copernicus's table of anomalies and additiosubtractions, we can now calculate the apparent movement of the sun; one column reads: "Additions-or-subtractions arising from eccentric orbital circle or first epicycle." He will stay a Ptolemaist to his death. But then he reminds us why we should lovingly remember him: "And so the explanation of the appearance of the sun *by means of the mobility of the Earth* is consonant with ancient and modern findings; and it is all the more presumed to hold true for the future."

Silent to the End

"A pale, insignificant figure"

"As a person," concludes Arthur Koestler, "he seems to be a pale, insignificant figure, a timid Canon in the God-forsaken Prussian province of Varmia; his main ambition ... was to be left alone and not to incur derision ..."

Jacob Bronowski writes: "His character remains silent for us to the end. He never married, though ... he had a housekeeper who was looked at askance as late as 1539."

In his portraits we sometimes meet him clinging to his great book, peering cautiously at us out of darkness. Often his cheekbones and other facial angles are pronounced so that, darkhaired and darkeyed, he almost resembles a Mongolian or Amerindian. He has been represented with and without praying hands.

"He eschewed all ordinary society," writes Sir Robert S. Ball, "restricting his intimacies to very grave and learned companions, and refusing to engage in conversation of any useless kind."

One of those companions, the young mathematician Rheticus, knew Copernicus personally and sums him up thus: "The astronomy of my lord and teacher can justly be called eternal."

Postludes to an occultation

His real name, which was also his copper-trading father's, happened to be Niklas Koppernigk, sometimes simplified Kopernik, and he came into this world in the riverine city of Thorn, or Torun if you prefer, on 19 February 1473. What good can this information possibly do us? To what eclipse or occultation can we link it? What have we to do with his noble Silesian mother or his other antecedents?

In a plan from the following century, Torun resembles a broken star with eight points left, its imperfect edge fronting the Vistula. In this respect Torun scarcely differs from the old universe after Copernicus finishes with it.

A thirteenth-century chronicle informs us that the city was originally a fort built around the trunk of a great oak tree by Landmeister Hermann Balk. The Vistula's frequent floods required the relocation of the place. Here, too, we might if we wished draw a parable about some old universe. But why trouble ourselves?

If we peered into the tall-walled house, great brick in a wall of other houses, where he was born, what could we discover? If we haled him before us now, what would we want to know? What would he tell us? Bronowski again: "When in doubt, he preferred to remain silent, and said nothing that he did not believe."

What *does* he believe?

We find no evidence that he ever became an ordained priest. All the same, thanks to his powerful uncle, he's been a Canon of the Catholic Church ever since age twenty-two. His father the copper-trader cannot help him, for he died in the same year when the Alphonsine Tables were finally printed; Copernicus was ten. This detail corresponds with the impression I've formed of my hero's retiring, isolated soul.

The stipend is comfortable and guaranteed until death. But he has to judge legal cases among peasantry, take in rents, inspect church farms, etcetera. We can't be certain how much leisure he possessed for his astronomical studies. In 1625, the scholar Strawolski saw fit to leave us this assessment of the man:

> In medicine he was renowned as the second Aesculapius . . . which should be interpreted that he had known some simple remedies, prepared them himself, and happily utilized, distributing them amongst the poor, who worshipped him therefore as if he were some deity.

If Strawolski tells that tale truly, then the tower, ruled stick and sheets of geometrical diagrams must have been very occasional diversions. But as usual, we don't know.

In about 1491, the eighteen-year-old Copernicus, now at the Jagiellonian University, begins to study law or astronomy, or it might be that he studies astronomy and law. (Cracow's densely crystalline aggregates of high roofs and sharp towers live within the many-towered, many-gated wall; a narrow bridge leads across the river to the lesser island called CASMIRVS, whose bridges cling like spider-legs across the motion of the aquatic sphere; so a contemporary plan depicts this two-membered constellation. In the stony courtyard-sea rises an archway'd island, surmounted with a cathedral cross: Jagiellonian University.) Very likely it is at this point in his life that he buys or is given the second printed edition of the Alphonsine Tables (Venice, 1492). Contrary to his usual practice, he orders blank pages bound in with his copy, perhaps so that he can bring it up to date with the new *Tabulae Resolutae.*

He is known to have read Sacrobosco's *On the Spheres* and Peurbach's *New Theory of Planetary Motion,* another Ptolemaic treatise. He is *thought* to have attended Wojciech Krypa's lectures on Ptolemy, which must have presented *The Almagest* as a series of verities nearly as true as Scriptures themselves.

In 1494, following the approved course of a young Polish gentleman, he explores foreign reaches of our sublunary world. Specifically, he attends the University of Bologna (whose cityscape is less heavy architecturally than Cracow, but still walled and courtyarded), and here he studies both law and astrology. Needless to say, the latter discipline requires its adherents to make detailed planetary observations. And so he quickly returns to his mind's true home, the superlunary realm. His teacher is a certain Domenico Maria de Novarra, with whom he observes the lunar occultation of Aldebaran in 1497. He discovers that the moon's parallax during that event refuses to correspond to Ptolemy's predictions.

Meanwhile he studies Greek, reading *The Almagest* in original. His doctorate in Ferrara, his other travels to and fro, what are those to us? It has been written that around this time he first begins to mull over Pythagoras's notion of a sun-centered cosmos, but why couldn't it have been earlier? We can see into his mind as well as he could see the disk of Mercury. But it seems likely that in these circuits of his he gets exposed to the fashionable ecstasies of the Neoplatonists, who preach that a finite Aristotelian universe would limit the perfection of God. (Remember *Revolutions*'s appeal to perfection, which we've already quoted as an example of Scriptural astronomy: "In the center of all rests the sun. For who would place this lamp of a very beautiful temple in another or better place than this wherefrom it can illuminate everything at the same time?") Although *Revolutions* will never intrude itself beyond the Sphere of Fixed Stars, that Neoplatonist instinct outwards might have Copernicanism's uncenteredness. Shall we then label him a Neoplatonist, as Kuhn does? His editor Rosen indignantly replies: "These groupings lean on a broken reed sliding on slippery mud, while ignoring Copernicus's firm attachment to the solid core of Aristotelianism." Well, well; after all, who knows Copernicus?

In 1500 he gives mathematics lectures in Rome. What else might he have done there? How much time could you imagine him to trifle away on courtesans and monuments? In *Revolutions* he relates the technical details of an eclipse observed ever so long ago by Ptolemy, then remarks: "We made careful observations of the other eclipse at Rome, in the year of Our Lord 1500, after the Nones of November, 2 hours after midnight, and it was the 8th daybreak before the Ides of November." So he chases his passion in Rome.

In 1503 he returns to Cracow to be secretary and private doctor to his uncle. We're informed that thenceforth he "led a solitary life devoted to a multitude of occupations."

In 1504 or thereabouts he observes a conjunction of all five planets, sun and moon in Cancer; their positions appear to be off from where the Alphonsine Tables claim they ought to be.

In about 1506, he begins working out the mathematics of his heliocentric system.

Sometime between 1508 and 1514 he writes his *Commentariolus*, which among other potentially heretical opinions opines:

1. "THERE IS NO ONE CENTER OF ALL THE CELESTIAL ORBS OR SPHERES."
3. "ALL THE SPHERES ENCIRCLE THE SUN ... SO THAT THE CENTER OF THE UNIVERSE IS NEAR THE SUN."
6. "WHAT APPEARS TO US AS MOTIONS OF THE SUN ARE DUE, NOT TO ITS MOTION, BUT TO THE MOTION OF THE EARTH AND OUR SPHERE, WITH WHICH WE REVOLVE AROUND THE SUN."

Speaking of heresy, we might note that thanks to the susceptibility of sixteenth-century Polish kings to their Protestant advisers, which prevents much of the sectarian slaughter which is occurring in other European countries, Poland gets nicknamed *Paradisus Hereticorum*, the Paradise of Heretics.

Fish days and meat days in Gynopolis

In 1510 he moves to Frauenburg, Frombork if you prefer, Citadel of Our Lady as we could literalize it, Gynopolis as he playfully Greekifies it, in hopes of more peace. I suspect that peace is what he finds. In 1511, 1522 and 1523 he observes the three total eclipses, measuring their parameters with his rough instruments. In 1512, 1518 and 1523, he attends the three solar oppositions of Mars. "We ourselves made observations of a second position of Venus in the year of Our Lord 1529 on the 4th day before the Ides of March, 1 hour after sunset ... and we had a full view at Frauenburg." The mathematician R. F. Matlak gives him credit for solving two then vexing difficulties of spherical trigonometry: how to find the angles when all sides are known; and vice versa. How could he have succeeded in those accomplishments if life didn't leave him alone?

To be sure, meanwhile he defends against epidemics and secures the water supply for Warmia (Varmia as others say). Alexander Rytel, M.D., believes him directly responsible for designing the water tower a hundred feet high which raised water on "two prismatic rollers"; a bucket and chain filled the reservoir. But this claim has been disputed by others.

An official reception is called for. "Venerable and worshipful gentlemen, honorable masters . . . ," he writes, "the arrangements are virtually complete for either" circumstance, "whether it happens to be a fish day or a meat day."

Meanwhile, in his hand, as an annotation to the work of Regiomontanus, we have the world's first table of secants.

Jacobsen believes him to have been "perfectly happy in his comfortable, if somewhat obscure, ecclesiastical position, where he enjoyed a reputation as a devout Roman Catholic, a fine and helpful physician, a good administrator, and an astronomer of the very first rank. He most certainly did not wish to be a martyr for anything." I myself can't help but wonder how high his aspirations were. I imagine him sincerely hoping to solve all celestial problems, to save the appearances and likewise to explain them. I see him as one of us, a man who lived on Earth and will not come again, a man whose dreams were greater than could be achieved, a lost one, enchanted by something beyond him, a man who gave his best to something, died, and left his accomplishment rusting into obsolescence. He was long ago and we cannot remember more than scraps of him. And this is what it means to be one of us in this sublunary world, a person whose hopes will all sooner or later be superseded.

In 1515, with his elder brother Andrew's leprosy now far advanced, he begins *Revolutions*, which he might have finished as early as 1530, but in the end he'll remark that "the scorn which I had reason to fear on account of the incomprehensibility and novelty of my theory had almost persuaded me to completely abandon the work which I had begun."

And yet he's not quite Koestler's dreamy mushroom man. Triangulating on his whereabouts, we find him in this argument or that controversy. In 1524, the venerable Heinrich Snellenberg procrastinates in paying Copernicus the ten marks he owes him. Our protagonist accordingly writes the Bishop of Varmia: "I see therefore that . . . my reward for affection is to be hated, and to be mocked for my complacency." And he respectfully requests that the Bishop withhold income from Snellenberg's benefice until the debt is made good. He's already proved himself capable of writing sarcastic cracks against his colleagues in sky-geometry: "Being a great astronomer, (Werner) is not aware that around the points of uniform motion . . . the stars' motion cannot possibly appear more uniform than elsewhere." Taking sides in favor of uniform circular motion, he passionately demands the repudiation of Ptolemy's equant: "For what else shall we be doing except giving a hold to those who detract from this art?" Why then would this tiger fail to stand up for heliocentrism?

In 1516 a memorandum mentions him as having given advice on calendar reform. On the other hand, we're also informed that in 1517 he chooses *not* to help reform the calendar, since solar and lunary motions remain inaccurately understood. It has been said that this matter of reform is what impelled him to work out the details of his heliocentric theory, but I find this difficult to credit, given the prior existence of the *Commentariolus* and the nipping wound he's already inflicted upon Ptolemy during that occultation of Aldebaran. Whatever the case, calendar reform remains a matter which *Revolutions* addresses directly and indirectly; for instance, part of Book IV's reason for being (it treats of lunar matters) is this: "As the year belongs to the sun, so the month belongs to the moon."

In 1520 his residence gets destroyed by war, a force eternally integral to our sublunary world. From 1521 on he remains at Frombork, waiting on this star or that occultation.

"Copernicus was a dedicated specialist," writes Kuhn. ". . . For him, mathematical and celestial detail came first; he wore blinders that kept his gaze focused upon the mathematical harmonies of the heavens."

Schnitt's world map of 1532 already shows terrestrial rotation. The cause, pictorially at least, is angel-power. But in 1533 the Pope receives word of Copernicus's theories through the scholar Widmanstal. Fortunately, the Pope proves indulgent. Well, after all, isn't the Copernican universe finite, bounded by the Sphere of Fixed Stars, superficially Aristotelian? Moreover, our middle-aged Canon's mathematical conceits may finally allow meat days and fish days to obey the calendar. Indeed, in 1535 Copernicus completes planetary tables of superior accuracy. The following year he sends his general almanac for publication to Vienna, although for some reason it does not get published after all.

In 1536 the twenty-two-year-old Rheticus, thanks to the patronage of Copernicus's future enemy Melanchthon, receives a professorship of mathematics at the University of Wittemberg. Three years later, having visited his "lord and teacher," he writes the summary of *Revolutions* entitled *Narratio Primo*. I'm told that it "saves the appearances" still better than Copernicus's *Commentariolus*. Some historians cite the *Narratio* as the first printed exposition of heliocentrism.

In *The Book Nobody Read* (and you can guess which book is being referred to), a historian suggests that "life's exigencies had prepared Rheticus to be a rebel"—for Rheticus's father had been decapitated for swindling, while he himself was Protestant and possibly homosexual—"and the heliocentric cosmology, so contrary to the deeply rooted beliefs of the day, must have inflamed his imagination." But how much of a rebel did one need to be in those days, to dabble in a book nobody read?

"Nobody shall have any proper pretext to suspect evil of me hereafter"

That's the question, isn't it? Does the premise of *Revolutions*, which will soon be called Copernicanism, require daring of its originator? How subversive is the recluse of Gynopolis?

In 1524 he writes to a reverend Canon in Cracow who happens to be Secretary to the King of Poland: "Faultfinding is of little use and scant profit . . . ," he advises us. "Hence I even fear that I may arouse anger if I reprove another while I myself produce nothing better." But this quietism need not be taken as anything more than rhetorical, for directly afterward he commences attacking Werner, the so-called "great astronomer."

In 1532, the dislikeable Johannes Dantiscus becomes Bishop of Chelmno and invites Copernicus to his installation. The latter writes back that he is "required by certain business" to be excused. "This refusal required courage on the part of Copernicus," an adoring commentator assures us, and very likely it does, for Dantiscus might easily someday become Bishop of Varmia, Copernicus's immediate superior.

Naturally, Copernicus bows to authority when he must. His former housekeeper, who in defiance of Copernicus's advice has separated from her husband, perhaps because he is impotent (we'll believe the commentator although the letter says something rather different), stays at Copernicus's home in Fromborg in 1531 with her new employer, who may or may not be a pious lady. All the same, two women under his roof at night, and one of them already the subject of rumors! No wonder that poor Copernicus gets a curt letter from the Bishop.

"But since I realize the bad opinion of me arising therefrom," he replies, "I shall so order my affairs that nobody will have any proper pretext to suspect evil of me hereafter, especially on account of your Most Reverend Lordship's admonition and exhortation."

In 1538 he is ordered to replace his new housekeeper with a female relative in order to avoid scandal. He asks for more time, which he is refused. "I acknowledge Your Reverend Lordship's quite fatherly, and more than father admonition, which I have felt even in my innermost being," he replies, and gets rid of his housekeeper in the stipulated time.

In the same spirit, he seasons *Revolutions* with such cautious expressions of submission as: "In the case of the other planets I shall try—with the help of God, without Whom we can do nothing—to make a more detailed inquiry concerning them . . ." Of course, caution comprises a mere part of the cause. These effusions are the fashion of the times, and very likely true belief—wasn't Copernicus a dignitary of the Catholic Church?

Safe at last

In 1541 Copernicus must still be up to earthly pursuits, since he doctors one of Duke Albrecht's courtiers. Two years later he commits himself to God's mercy, with the first published copy of *Revolutions* in his dying clutch. What is any life? The sun traverses some degrees of a circle, and then it is night.

As for his legacy, let's go bobbing for apples once more in Koestler's barrel of compliments: "It is one of the dreariest and most unreadable books that made history."

Exegesis

Book IV

" Since in the preceding book, to the extent that our mediocrity was able, we explained the appearances due to the movement of the Earth around the sun . . ." Thus begins Book IV, whose subject will be the moon. "Through her in particular, who shares in both night and day, the positions of the stars are apprehended." Moreover, "she alone of all the planets refers her revolutions however irregular directly to the center of the Earth and is most closely akin to the Earth"—a fact which fed our delusion of being the center of everything else.

Copernicus's theory now grows complicated because the lunar appearances are complicated. "For the moon is changeable even from hour to hour and does not stay like itself."

IV.2–4: "I say that the lunar appearances agree"

He determines the lunar movement by reckoning up Ptolemy's observations of eclipses and his own: "Now it is clear that in the middle space of time between the first and the second eclipse the moon traversed as much space as the sun in its apparent movement—not counting the full circles—*i.e.,* 161° 55', and between the second and the third eclipse, 138° 55'." In the nearly 1389 Egyptian years of this interval, Ptolemy's calculated lunar movement away from the sun is off by twenty-six minutes; his calculated movement of anomaly is off by thirty-eight minutes. Copernicus proudly reports that his own numbers are consonant with the appearances.

The lunar orbit is canted to the ecliptic at its own peculiar angle. Copernicus will blame part of the moon's apparent irregularity on its obliquity (whose value we now calculate at 5° 8' 43")—for his theory forbears to demand that all circles lie in the same plane. Indeed, as has already been noted, the planes of Copernican planetary deferents and their epicycles, which lie parallel to them, are all permitted to tilt off Earth's orbital plane. Luna accordingly "bisects the ecliptic and is in turn bisected by it, and from this line of intersection the moon crosses over into both latitudes."

The moon thus "revolves obliquely around the center of the Earth in a regular movement of approximately 3 minutes" of arc "per day, and it completes its revolution in 19 years. That is, every 18 years 223 days, a specific lunar phase will repeat *and* the moon will be in its starting position relative to the ecliptic. How might Copernicus's predecessors have worked this out? Well, a celestial drama called their attention to it: It is the length of time over which the pattern of lunar eclipses repeats.

More observational grist for Copernicus's logic-mill: The moon seems to pass by more quickly when it is closer to us.

"Accordingly the ancients understood that change in velocity to occur on account of an epicycle; in running around this epicycle the moon, when in the upper semicircle, subtracts from the regular movement, but when in the lower semicircle, it adds the same amount to it."

And Copernicus draws a diagram of what "the ancients"— who begin to remind me of a certain Claudius Ptolemy— must have had in mind. "But if this is so, what shall we reply to the axiom: *The movement of the heavenly bodies is regular except for seeming irregular with respect to the appearances;* if the apparent regular movement of the epicycle is really irregular . . . ?" What Copernicus finds unacceptable turns out to be, as you might have guessed, the loathsome equant, which here gets insinuatingly described as "some other different point, which has the Earth midway between it and the center of the eccentric circle."

His own solution to the apparent lunar irregularity remains loyal to first principles: Each month the moon loops twice around an epicycle which in turn gets carried around a larger epicycle which during the same interval "makes one revolution with respect to the mean position of the sun." The radius of the small epicycle is 474 units; the center of that circle describes a circle of radius 1097 units, all determined through geometric "additiosubtractions." "I say that the lunar appearances agree with this setup."

Do they? Here is how our own astronomers see it: "The orbit of the moon around the Earth is approximately an ellipse." Indeed, the mean lunar eccentricity is substantial: 0.0549. The Earth-moon distance can vary by as much as 14 percent, or 51,000 kilometers.

IV.4–32: Distances, diameters, volumes

He tells how to compute the parallaxes of the sun and the moon with the aid of his tables—a miserable series of geometric and arithmetical operations which you will thank me to omit. He shows us how to calculate how great an eclipse of the sun or moon will be: Get the latitude at the moment of conjunction, subtract it from half the diameter of the celestial orb in question, multiply that by twelve and divide the product by the diameter of the orb, at which point "we shall have the number of twelfths of the eclipse." He also provides a formula for determining the duration of any future eclipse.

After discussing various parallaxes of the moon, pointing out each time how Ptolemy's science is not quite right in this regard, he announces: "From this it will now be apparent how great the distance of the moon from the Earth is. And without this distance a sure ratio cannot be given for the parallaxes, for they are mutually related." The figure he obtains is 56 times the Earth's radius plus 42 minutes of arc—in other words, 56.7. Our current value is 60.27 times the Earth's equatorial radius, which I am informed is "within a few percent" of the value obtained by Hipparchus seven centuries before Copernicus. So Copernicus's calculated lunar-terrestrial distance, while creditable, is no improvement over that of the ancients. Indeed, it falls short of Ptolemy's mean value of 59.

He also obtains the diameter of the moon: "The comparison of the difference in extent of the eclipses with the latitude of the moon shows how much of the circle around the center of the Earth the diameter of the moon subtends. When that has been perceived, the semidiameter of the shadow is also known." Copernicus's figure, "in accordance with Ptolemy's conclusion," for the lunar diameter is 31' 20". This is very close to our own value for the (apparent mean) diameter of 31' 52". One of our uncentered era's not wildly pro-Copernican specialists will praise him for "improving" the angular diameter of the moon at perigee from Ptolemy's value of almost a full degree to 37' 34"; the modern value is 33' 32".

In connection with the diameter of this perfect Aristotelian disk we had better repeat that the moon, like our poor sad Earth, is aspherical; one lunar axis exceeds the other by not quite three kilometers, which is significant enough to cause the moon to keep one side eternally facing Earth. Knowing this might have broken Copernicus's heart.

By a similar demonstration he arrives at the solar-terrestrial distance at apogee: 1179 times Earth's radius. Here he does not do well; the actual value is 23,455.

He computes that the sun is 161 and 7/8 times greater in volume than the Earth, which is in turn 42 and 7/8 times greater than the moon. "And hence the sun will be 6,999 and 62/63 greater than the moon."

His ratio of terrestrial to lunar volumes is not so far from our own figure: 50 to 1. The solar volume, however, he grossly underestimates; it is actually 1,306,000 times greater than Earth's, 653,000,000 times greater than the moon's. Hence for the ratio of solar to lunar volumes he is off by a factor of more than 9000.

The Pillars of Hercules

That outstanding poet of Scripturalist astronomy, Dante, places Earth where it should be, at the center of the universe, and within our home orb, in descendingly concentric spheres of ascending torment, sinners get compartmentalized. In the Eighth Circle, the second most horrid, Dante ranks ten sorts of fraudulent souls; and among the eighth worst of those we find Ulysses and his companion Diomed getting tortured for all time in a horn of flame. What was the sin? They'd passed beyond the Pillars of Hercules, limits of the world! "Choose not to deny experience," Ulysses exhorted his sailors. "You were not born to live as brutes, but to follow virtue and knowledge." Thus he became a false counsellor. From the forbidden mountain they glimpsed ahead (Purgatory), came a storm which, "as One willed," overwhelmed them. A commentator writes that "here, if anywhere, Dante's imagination beats at the bars of his day and creed."

(Fellow upholders of the good old universe, you will be reassured to learn that Dante matches up his universe's three movements—epicyclic revolution, solar and heavenly revolution, and precession—with the three angelic Thrones).

"I doubt not that certain savants have taken great offense"

What about Copernicus? Does he sense that *Revolutions* has brought us all to the Pillars of Hercules?

In his *History of Polish Literature*, Milosz opines that Copernicus "was not eager to publish, as he feared to provoke a scandal."

All the same, when his opus does finally enter the light, it is Osiander's degrading preface, not any word by Copernicus, which sounds a defensive note:

> Since the newness of the hypotheses of this work, which launches the Earth into motion and places an immovable sun at the center of the universe, has already received considerable publicity, I doubt not that certain savants have taken grave offense . . . If, however, they are willing to weigh the matter scrupulously, they will discover that the author of this work has done nothing deserving of blame.

Copernicus sticks to his customary strategy of pretending that his novelties have ancient sanction, as they most certainly do. Do you remember what the Council of Trent determined? "It should be forbidden to interpret the Scripture contrary to the declared sentiment of the Church, and to the unanimous consent of the Fathers . . ." Wouldn't it be lovely if our Catholic traditions could be proven to be in consonance with Pythagorean fantasies? Well, why not? Copernicus is still safe, not only since he dwells in torpid Gynopolis, but more relevantly because the Church has not yet come out explicitly against heliocentrism. In fact, various bishops support it! For that matter, we're informed that in the ten years before *Revolutions* was printed, Pope Clement VII not only did not stop his secretary, Johann Albrecht von Widmanstetter, from lecturing on Copernicanism in the Vatican garden, but afterwards even rewarded him with a Greek manuscript.

"By the second decade of the seventeenth century," writes Kuhn, "Catholic authorities were giving greater weight to scriptural evidence and allowing less latitude for speculative dissent than they had done for centuries." Fortunately, we're but midway through the sixteenth! And Copernicus has protected himself further by means of his dedication to the Pope. I myself wouldn't be surprised if the attacks upon him by Protestants, who in order to better reject Catholic authority must insist on Scripturalist literalism (the Earth stands still because the Bible says it does), endear him all the more to his own.

"As for the Catholic hierarchies," Santillana tells the tale, "they held Copernicus in respect as a church man and a scholar, but they considered his system as one of those ingenious mathematical devices which could lay no claim to physical reality. Mathematics was rated at the time as the thing for technicians and *virtuosi*, as they were called, with no claim to philosophical relevance . . ."

In short, Copernicus has made his life-voyage under peculiar atmospheric conditions. Poland's skies are hazy, and even the alert stargazers of the Church cannot see far enough beyond them to descry, dark and void within the void dark night, the silhouettes of the Pillars of Hercules, where he has come, perhaps despite his own inclinations; he's sailed obscurely and securely beyond the limits of observation . . .

To the Eighth Circle

Shall we say, then, that Copernicus died in his bed instead of at the stake because he was lucky enough to be one of the *virtuosi*? He "was not eager to publish, as he feared to provoke a scandal," but what danger did he actually run? Haven't we just agreed that he was protected by his own arcaneness?

Pope Clement attended a Copernican demonstration in the Vatican garden; that's true; but Pope Clement is dead. It is likewise the case that Copernicus dedicated *Revolutions* to Pope Paul II, but Paul never did approve it as far as we know. The Pope's deputy, Cardinal Schönbeg or Schönberg, sometimes receives mention in this regard; but apparently only one letter from him remains extant, a courteous enough missive requesting to see the manuscript of the still unfinished *Revolutions*. Schönbeg died ten months later, in September 1534, so it is unlikely that there was time for him to receive whatever portions of *Revolutions* had been completed, read them, show them to the Pope, and pass on to Copernicus the glad news of the Vatican's full and unlikely approval.

In fact, Copernicus's situation was precarious. Do you remember Father Giovanni Maria Tolosani? He's the Dominican we've quoted against Copernicus on several occasions: *Revolutions* disobeys Aristotle's laws of motion, places the sun in an inappropriate place, loses sight of the Empyrean Heaven! All the same, Tolosani must not have dipped too far into *Revolutions*. He might have been misled by Osiander's preface, for here is what he has to say about heliocentrism: "Nobody accepts it now except Copernicus. In my judgment, he does not

regard that belief to be true." (*Revolutions*, I.5: "*The case is so.*")
And if Copernicus does not believe that the Earth moves, why
not let him scribble his foolishness in peace?

But Father Tolosani now goes on to say, more threaten-
ingly, that Copernicus "has no right to complain about the
men with whom he disputed in Rome, and by whom he was
most severely condemned."

Condemned! Whatever can this mean? As we know, Copernicus has not been in Rome since his youth, when *Revolutions*
and the *Commentariolus* both remained unwritten, and their
author was observing eclipses and such. In the opinion of Rosen,
to whom we are indebted for this story, Father Tolosani must
be referring to Copernicus's defender and fellow Canon of
Varmia, Alexander Scultetus, who voyages to Rome and publishes a book there in 1546, three years after Copernicus's
death. (Tolosani himself will die in 1549.) Scultetus's book
speaks approvingly of the Copernican doctrine.

We're informed that Tolosani is very close to the Vatican
through his friendship with one Bartolomeo Spina of Pisa, who
has been appointed Master of the Sacred and Apostolic Palace.

With these details in mind, let us consider one more
extract from Father Tolosani's contribution to the Copernican debate, which is entitled *On the Truth of Holy Scripture*.
He is speaking of *Revolutions*:

> The Master of the Sacred and Apostolic Palace had planned
> to condemn his book. But, prevented first by illness, then
> by death, he could not carry out this. This I took care to
> accomplish afterwards . . . for the purpose of safeguarding
> the truth to the general advantage of the Holy Church.

Copernicus, then, was not only lucky, but *canny* to have died when he published. They've doomed him, but too late; he's already beyond them, on the far side of the Pillars of Hercules.

Herschel's looming universe

So the Church knew. But let's ask one more time: Did *he* know how far he'd sailed?

H. G. Wells, whose science fictions epitomize human beings lost in one natural immensity or another, tells *Revolutions*'s tale succinctly: "The globe of Earth was the centre of being; the sun, the moon, the planets, the fixed stars, moved about it as their center, in crystalline spheres. It was only in the fifteenth century [*sic*] that men's minds moved beyond this, and Copernicus made his amazing guess that the sun was the center and not the Earth."

In spite of that amazing guess, which was amazing, it's true, but not a guess, Copernicus, like Aristotle, Ptolemy and Father Tolosani, cannot let go of the *smallness* of the old universe which he's demolishing.

Here I may be projecting my own hollowed uncenteredness upon the man. Consider the opening sentence of Book VI: "We have indicated to the best of our ability what power and effect the assumption of the revolution of the Earth has in the case of the apparent movement in longitude of the wandering stars and what a sure and necessary order it places all the appearances." Copernicus is pridefully—and rightfully—harking back to geocentrism as the antipode of comparison; whereas I, corrupted by Keplerism and undone by Newtonianism, hark forward to gravitation, which reduces the Earth's revolution to near irrelevance, as the cause of the *true* movement of those

wandering stars. "All the appearances," he says! But in my day
the limits of observation have been pushed much farther away
into the blackness; ever so many intergalactic appearances are
what they are without measurable reference to the Earth's rotation. There you have my reaction—a misjudgement of poor
Copernicus, who had such "appearances" as planetary retrogression in mind. Just the same, he did say "all the appearances." He might have thought he'd sailed far enough to see it
all; and what he saw remained within the known "world."

Again and again he refers to "the proper annual movements
in relation to the Sphere of the Fixed Stars." What if there were
no Sphere of the Fixed Stars? "The mind shudders."

In his discussion of Venus and Mercury, he accepts the
necessity which his predecessors felt to fill the interplanetary
void: "In order for such a vast space not to remain empty, they
find that the intervals between the perigees and apogees—
according to which they reason out the thicknesses of the
spheres—add up to approximately the same sum" as the sums
of the waste distances. "We do not know," he drily remarks,
"that this great space contains anything except air, or, if you
prefer, what they call the fiery element." And he attacks
Ptolemaism on that specific point: The immense epicycles
and equants required to save the appearances require incomprehensible volumes! Hence that rhetorical cry which I have
already quoted: "Then what will they say is contained in all
this space?" (Nietzsche, *ca.* 1886: "Since Copernicus man has
been rolling from the center toward X.")

Yes, Copernicus has voyaged beyond his own limits of
observation, not daring to see (how can our hearts not go out
to him in his blindness?) that the forbidden entity which lies
beyond the Pillars of Hercules is Infinity.

Exegesis

Book V

"Now we are turning to the movements of the five wan-dering stars," pursues Copernicus; and knowing as we do his attachment to circular uniform motion, we should not be surprised that the first part has been entitled "On Their Revolutions and Mean Movements."

Such celestial twirlings he quite properly subdivides into two, those which are "proper to each planet," and the various apparent pausings, retrogressions and progressions which occur in an Earthly observer's frame of reference "by reason," as Copernicus explains, "of the parallax caused by the move-ments of the Earth taken into relation to the differing magni-tude of their orbital circles."

V.1–5: The Martian circles

Firstly, consider retrogressions. Specifically, consider the case of Mars. Almost twice per year, the red planet (which to Copernicus would have been merely a yellowish star) seems to back water against the night, then resumes its course at a lesser height than before, so that its track resembles an S-curve whose beginning and end both pull outward infinitely. How to explain this? Pity Ptolemy, who, clinging to the false simplicity of geocentrism, invested his mathematics in an eccentric whose center traveled eastward around the anthropomorphically mis-defined ecliptic's center at a speed equal to the sun's misconstrued velocity about Earth, while upon this eccentric, Mars traveled westward "at a speed equal to the anomalistic passage, and if a straight line is drawn to the eccentric circle . . . through our eye . . . in such a way that one-half of" it "has to the lesser of the segments produced by our eye the ratio which the eccentric's speed has to the star's speed," then that star, Mars, will sometimes seem to retrogress!

And Copernicus? Steadfastly he continues to show us the solar system from the point of view of an uncentered Earth, which renders the planetary retrocessions as accidental and arbitrary as the stations of Copernicus's own life—Padua, Ferrara, Lidzbark, Olsztyn, Frombork. Specifically, he reminds us that prior geometers have explained the retrocessions of planets as being due only to the movements of those planets in relation to the sun—which are really caused by "the parallax due to the great orbital circle of the Earth." (The modern long and short of it is Mars travels more slowly round the sun than we do, and at some slight obliquity to the ecliptic.) And Copernicus triumphantly announces (the italics are his own; I omit the geometric proof): "I say that *when the planet is set up at point F, it will present to us the appearance of stopping; and that whatever size of the arc we take on either side of F, we shall find the planet progressing, if the arc is taken in the direction of the apogee, and retrograding, if in the direction of the perigee.*"

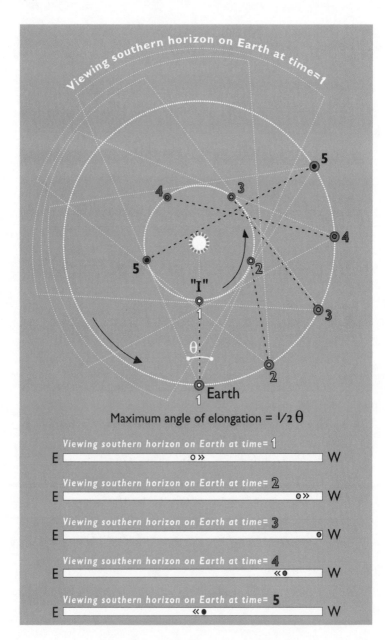

Viewing southern horizon on Earth at time=1

"I"

1

θ

Earth
1

Maximum angle of elongation = ½ θ

Viewing southern horizon on Earth at time= 1

E _____ o ≫ _____ W

Viewing southern horizon on Earth at time= 2

E _____ o ≫ ____ W

Viewing southern horizon on Earth at time= 3

E _____ o W

Viewing southern horizon on Earth at time= 4

E _____ ≪ o ____ W

Viewing southern horizon on Earth at time= 5

E _____ ≪ ● ____ W

Figure 18 *Apparent Retrograde Motion of an Inner Planet I (Simplified Copernican View)*

For each interval, turn the diagram and imagine a viewing horizon comparable to 1, with Earth at bottom center.

Apparent position, travel direction and velocity of I as projected on our horizon (assuming that we could always see it):

In this generalized diagram, the inner planet's relative velocity is exaggerated, and its apparent motion simplified. (Of course the appearances are knottier. Consider the apparent path of Venus as seen from the Earth: In the course of two apparent circuits around us, whose three benchmark points are not at all the same, Venus executed one retrogression [the loop at R].)

Assuming uniform counterclockwise circular motion for both Earth and I, and also assuming that I travels faster than our planet, if I and Earth are at inferior conjunction at time interval 1, I will be seen to be in the middle of our horizon. It will then draw farther and farther toward the west. As Copernicus notes, it will move westward with relative slowness, since both planets are going in the same direction, so that some of I's perceived motion gets cancelled out by Earth's.

At $t = 3$, I will have moved as far west as it can upon our horizon, since it is near its maximum angular elongation.

At $t = 4$ and $t = 5$, I has already reached the far side of the sun. From our point of view it is now moving eastward. Since Earth is still moving the same direction as before, I's apparent speed across our horizon increases.

He need not build his new theory of the planetary motions from scratch, for Ptolemy's geometric calculations, whose goal was correspondence with observation, did express true mathematical relationships. In a long footnote, *The Almagest*'s translator calls this theory of the retrogressions "almost equivalent to the conversion theorem by which one passes from the Ptolemaic theory to the Copernican theory of the outer planets. The ratio of the radius of the circle the eccentric's center moves on to the eccentric's radius" in Ptolemy "is the same as that of the epicycle's radius to the radius of the deferent in the epicyclic theory" of Copernicus. Suffice it to recapitulate, as should be intuitively plausible, that the center of Ptolemy's eccentric corresponds to Copernicus's mean sun.

And so *Revolutions* has carried us closer to apprehending the reality of the Martian orbit. Now, where *is* Mars? In our discussion of Venus's orbit we asked this question in a much more general sense: In which order should the Sphere of Venus be ranked? Having answered this issue, we now would like to be able to locate the coordinates of each planet at any time.

The first step will be to determine the periods of the planets.

Copernicus rightly says that "the true position of Saturn, Jupiter and Mars becomes visible to us only" during opposition, which, logically enough, happens "in the middle of their retrogradations. For at that time they fall on a straight line with the mean position of the sun, and lay aside their parallax." (As he has already told us, the inner planets cannot be seen at such times, so he takes their angles of greatest western and eastern elongation, then calculates the mean between them.)

For Saturn, Jupiter and Mars, he compares three "modern observations"—his own—with three ancient ones, thereby determining how long it takes for each planet to cycle around from a certain position between the sun and a specific fixed star back to that same position. He calls this movement "a single circle of parallax." This data will allow him to compute the planetary distances better than anyone before him. In his own words: "Let us pass over in silence the multitude, complication and boredom of the calculations."

On second thought, let's remark on those calulations.

"The standard procedure involved," begins an astronomical exposition from our own uncentered epoch (the text is merely half a century old), "consists in studying the motion of a body in the solar system, constructing an ephemeris based upon a gravitational theory and making a comparison between the ephemeris and observations."

Copernicus has his ephemeris, an ephemeris prepared for him by the selfless labors of Ptolemy, Hipparchus and others, but please remind yourself again to the end of his days he will lack a gravitational theory, whose function is to give the planet in question "its heliocentric motion, which is determined by orbital elements and the adopted masses of the perturbing planets." Please remember this solitary Canon in his Frombork Tower, patiently working out the planetary circles and positions, and doing it without what we now consider the essential tools of Newtonian mechanics. That's why his geometries grow as elaborate as the spider-starflowers of inscribed lines in Polish church-naves, and still he isn't done.

"The problem is complicated," continue our twentieth-century astronomers, "by the circumstance that observations are made from an observatory on the surface of the Earth. The calculation of a geocentric ephemeris thus introduces the orbital motion of the Earth about the sun into the problem . . . allowance must be made for the *lunar inequality* produced by the motion of the center of the Earth about the center of mass of the Earth-Moon system. This orbit is a reflection in miniature of the motion of the Moon about the Earth." Again, Copernicus will never possess any opportunity of allowing for the lunar inequality since he has no theory of gravitation to show him its existence and importance.

Given these tremendous difficulties, it is more than remarkable that the deferent radii which Copernicus calculates for the planets, which translate into the mean radii of their actual elliptical orbits, will be fairly accurate for Mercury and Saturn, the two bookends of the known planets (here his error exceeds 3 percent); in the case of Venus, Mars and Jupiter, his error will be 0.55 percent or less.

"Allowance must further be made for the fact that the observer moves about the axis of rotation of the Earth. This is done by applying a correction for the geocentric parallax to the observed position, thereby reducing it to a geocentric position. This correction requires the introduction of the solar parallax, and can be made if the distance of the observed body, expressed in astronomical units, is sufficiently well known."

This Copernicus can do—but only because he has worked out that "distance in astronomical units" in his own way: orbits of an incorrect shape about an incorrect point, the whole animated by an Aristotelian misunderstanding of motion! The astronomer Zdeněk Kopal accordingly assesses him as having labored "without much success."

Imagine Copernicus in the small hours of a Polish night, sighting through his astrolabe. "Moreover, we took observations on the conjunction of Mars with the first bright star of the Chelae—called the Southern Claw . . ." Drawing an arrow-pierced eccentric circle *ABC* connected by two elongated triangles to epicycle *BF*, Copernicus determines the mean orbital circle of Mars.

"We have compared these three of Ptolemy's observations of Mars with three other observations, which we did not take carelessly," he says in one place; and in another he computes an angle *LEM* of 47° 50', "whereby the movement of the planet from the first solar opposition to the second is apparent, and the number is consonant with experience." In brief, he determines the years between oppositions, transforms time into degrees of a circle in his accustomed fashion, and plots the position of each planet against the Sphere of the Fixed Stars. Then he draws his circles. "The known arc *AB* subtends the unknown angle *AEB* . . ." And he presently determines the mean movement of each planet's parallax.

In the case of Mars, *Revolutions*, relying on the observations of Hipparchus and Ptolemy, reports 37 "revolutions of parallax" in slightly more than 79 solar years, meaning that Mars gets overtaken by the Earth 37 times in this period, during which "the planet by its own movement completes 42 periods plus 2° 24' 56"." From this data, Copernicus (as often, he omits the details, but I presume that he divides 79 by 37, then multiplies by 365) derives a single circuit of Martian parallax of 779 days.

We now calculate the orbital period of Mars at 1.88 years, or 686 days. The value given by Copernicus of 42 periods in 79 years works out to 1.88, which confirms my admiration for the ancient observers—and for Copernicus; Ptolemy could have made this simple computation, had he only postulated heliocentrism first.

Another calculation from the figure for Mars's revolutions of parallax reinforces what has just been derived; for if Earth passes Mars 37 times in the same amount of time that Mars circles the sun 42 times, we should expect the Martian period to be 1 and 37/42 solar years, which again works out to 1.88 years.

"Moreover, we took observations on the conjunction of Mars with the first bright star of the Chelae—called the Southern Claw . . ." Drawing an arrow-pierced eccentric circle *ABC* connected by two elongated triangles to epicycle *BF*, Copernicus determines that the mean orbital circle of Mars will be equal to the ecliptic radius plus 31' 11".

The modern value for Mars's orbit assigns a mean radius of 1.524 astronomical units, or one and a half times the radius of the ecliptic. Dividing Copernicus's 31 minutes into 60 yields 0.516; add this to one ecliptic radius unit, or 1 AU if you prefer, and we obtain the close reach to reality of 1.516, all done with circles and angles from ancient observations! "So too in the case of Mars the movements, magnitudes and distances have been explicated in a fixed ratio by means of the movement of the Earth."

V.4–36: Rescuing Mercury from injury and disparagement

We now understand that the "proper movements" of each planet get shaped not only by its eccentricity, but also by its obliquity to the ecliptic, not to mention the inclination of its equator to its own orbital plane. Pluto, off kilter in so many other respects as well, boasts an inclination of 122° 28', while Venus's inclination is even higher: 177° 18'. Pluto possesses the highest obliquity (17° 8'), followed by Mercury at just over 7°. If only they all had the equatorial inclination of Mercury and the obliquity of Earth—pure zero! How Aristotelian, nay, how

Scriptural that would be! But we had better soldier on toward our uncentered universe of canted detritus.

Regarding our sort of eccentricity, namely deviation from a circle, Copernicus has certainly made his own view clear—no deviation permitted!—for he can smooth away irregularity with the other kind of eccentricity: an off-center circle. Should he employ a circle eccentric to an eccentric circle, or an eccentric circle which bears an epicycle? Either one sounds awfully beguiling to our hero. "By this composite movement," he sadly writes, "the planet does not describe a perfect circle in accordance with the theory of ancient mathematicians but a curve differing imperceptibly from one." No matter. "We shall demonstrate from observations that these hypotheses are sufficient for the appearances."

He saves Mercury from the equant, I mean "from liability to injury and disparagement," clothing it in an eccentric circle around an eccentric circle in place of the noxious equant. "When the figure has been drawn in this way, all these things will fall in order on the straight line *AHCEDFKILB*." Meanwhile he assigns to Mercury a period of 88 days, which is really not so far from our own value of 87.969 days.

What next? "Therefore by means of the tables drawn up in this way by us we shall calculate without any difficulty the positions in longitude of the five wandering stars." Including terrestrial motions within his purview brings him that much closer to what we might as well call absolute Truth; perhaps it is this consonance with reality which allows him to save Ptolemy's epicycles, at least for the outer planets, by translating them into circular orbits about the mean sun. (As for the eccentric circles of those planets, they become their mean longitudinal orbits about the mean sun.)

He applies his Earth-centered geometry to the position of Saturn for pages and pages. (Centuries later, a specialist will be moderately impressed by his "ingenious, though not successful, account of the heliocentric and geocentric latitudes of the planets.") Saturn's astrological image is of an old man, often wearing black and holding a sickle or crooked staff. I suppose this conception has to do with the great length of the Saturnian year. From a standpoint of the planet's distance from our life-giving sun, it is poetically accurate. Saturn swims far away in the cold blackness, yet not so far as Pluto, whose name and darkness we associate with death.

Harvesting Ptolemy's data, Copernicus notes that once upon a certain day in 1514, Saturn lay in a straight line with the stars in Scorpio's forehead. Marshalling parallax, mean solar position, the eccentric circle ABC and the ratio of triangles, he works out the greatest and least distances of Saturn. He does the same for Jupiter, adding: "All these things are in perfect regularity with our hypothesis of the mobility of the Earth and absolute regularity" of movement. And he sits alone at Frombork, while above him the planets sweep round and round to the east.

Assessments

W hat epitaph shall we give Copernicus?

"Rotting in a coffer"

To Koestler, who took part in revolutionary movements and therefore should have known better, the hero of this book was "neither an original nor even a progressive thinker" but a mere revisionist of Aristotle and of Aristotle's own more mathematical, less element-oriented revisionist, Ptolemy. *Revolutions* should be considered "one last attempt to patch up an outdated machinery by reversing the arrangement of its wheels."

Kuhn, who is both more fair and more kind, remarks that "Copernicus tried to design an essentially Aristotelian system around a moving Earth, but he failed. His followers saw the full consequences of his innovation, and the entire Aristotelian structure crumbled."

Aaboe dismisses *Revolutions* as "very much like *The Almagest* except in two respects. It includes cosmology, and it adheres strictly to the principle of uniform circular motion."

According to Santillana, "The great treatise of Copernicus had been known for a half-century, but in all this time it had aroused mostly skepticism. A few romantic and daring spirits had been captured by the new idea, but they could not master the difficult details of the system."

In the town of Elblag, however, Copernicus was evidently considered in his true threatening light, for as early as 1533 the Protestants had already composed a masque against him, entitled "A Stupid Stage." "He was presented as a haughty, cold, and aloof man who not only dabbled in astrology and considered himself inspired by God, but was rumored to have written a large work which was rotting in a coffer."

"Actually," laughs a Vatican astronomer in A.D. 2000, "the Copernican system, as Copernicus himself presented it, is only marginally simpler than the Ptolemaic system . . . Kepler's insight that orbits are elliptical was nearly a hundred years in the future." To be sure, *Revolutions* was "highly esteemed for the next 50 years . . ."

A Fellow of the Royal Aeronautical Society sums up astronomical developments as follows: "The planets, the 'wandering stars', gave the most problems of understanding. In the sixteenth century Kepler sorted them out." Poor Copernicus is not even mentioned. Let *Revolutions* rot in its coffer!

False supposition, true demonstration

"Kepler sorted them out." Well, what did Kepler himself say? Although he too lacked Newton's conception of gravity, he sensed all the same that the central sun somehow impelled the revolutions of the planets. He baldly stated that "there are no other smaller circles called epicycles." In addition to altering the Copernican orbits from circles to ellipses, he rationalized the planetary motions somewhat by asserting (as will be discussed below) that the radius from planet to sun sweeps equal areas in equal times—which is certainly a great advance over the Ptolemaic strategem of equants. All the same, Kepler's most beautiful treatise was entitled *The Epitome of Copernican Astronomy*, and he referred reverently (the capitals are his) to "the Philosophy of Copernicus."

In 1594 an English astronomer remarks on Copernicus's peculiar sun-centered notion, "by help of which false supposition he hath made truer demonstrations of the motions and revolutions of the celestial spheres than were ever made before."

The twentieth-century astronomer A. C. B. Lovell calls him "basically correct," and in the middle of the nineteenth century, our old acquaintance Herschel, himself a great scientific discoverer, writes simply in his *Outlines of Astronomy*: "We shall take for granted, from the outset, the Copernican system of the world."

Exegesis

Book VI

"It remains for us to occupy ourselves with the movements of the planets by which they digress in latitude," Copernicus begins this last book, "and to show how in this case too the selfsame mobility of the Earth exercises its command and prescribes laws for them here also."

Nowadays astronomy endeavors to describe for us the chemical compositions, surface temperatures, appearances, even topographies of our neighboring heavenly orbs; for their orbits have been worked out. We're informed that this or that planet orbits at such and such a mean distance from the sun, with so much eccentricity. The information appears as a line in a table, or as a subordinate clause in an introductory sentence. It's hardly news. As I wrote this book on my laptop computer, I enjoyed the luxury of switching back and forth between my text and an astronomy program which allowed me to watch the revolving planets from any vantage point I liked, with or without constellation labels and an ecliptic. Round and round went Earth; wiggle-squiggle went Mars as seen from Earth. It is the glory of Copernicanism to have brought us closer to this point. In the words of Ptolemy's translator: "Once the Copernican system is supposed, it is immediately possible to deduce from the Ptolemaic numbers"—circles of parallax, revolutions of longitude, radius of epicycle, etcetera—"*and without further observations,* two things which are extremely important for Kepler and Newton: (1) the periodic times of the planets about the sun and (2) the relative distances of the planets from the sun."

This is what *Revolutions* accomplished in Book V.

To Book VI, which is really a brief addendum to Book V, one chore remains: The planets' "true positions are said to be known only when their longitude together with their latitude in relation to the ecliptic has been established," he explains.

The greatest single failure of my own presentation of *Revolutions* is in the treatment of heavenly positions. What I should have done was to somehow find space to make real for you the construction of astrolabes and parallacticons, describing the movements of each instrument, the view through the eyepieces, then detailing the use of Copernicus's tables so that the transformation of observational data into positional knowledge would have been thoroughly understood. Unfortunately, every time I tried to do this, I expended twice as many words as Copernicus himself. In spite of my naive hopes, this book is therefore less an explication of *Revolutions* than a discussion of the issues which its reading engenders. An ideal companion volume to *Uncentering the Earth* would be written by a professional astronomer, and it would be entitled *How to Build and Use Your Very Own Back Yard Parallacticon, Complete with All Tables, Newly Updated with an Ephemeris for the Trans-Saturnian Planets.* Then you could better judge the merits of Copernicus's boast: "Accordingly by means of the assumption of the mobility of the Earth we shall do"—and only the next four words stir my skepticism—"with perhaps greater compactness and more becomingly what the ancient mathematicians thought to have demonstrated by means of the immobility of the Earth."

VI.1–8: Inclination, obliquation, deviation

In Book V Copernicus showed us how to calculate "the positions in longitude of the five wandering stars," meaning the distances of those planets westward from the mean solar position; and I spared you his wearisome scheme for arriving at the corrected parallax of a planet, which must be added to the corrected anomaly of parallax if it is "greater than a semicircle," and subtracted otherwise. The resulting planetary positions, when subtracted from the mean solar position, will give "the sought position of the planet in the Sphere of Fixed Stars." And in Book V, Copernicus calculated those for us.

He now turns to those apparent irregularities which are the result not of those differences in position and velocity which occur upon the ecliptic plane between the planet in question and the Earthly observer, but of the planet's obliquity to the ecliptic.

In his discussion of lunar movements in Book IV, Copernicus introduced the concept of **nodes**; but I interred it in a footnote since my summary at this point was superficial. Giving due credit to Ptolemy, he brings it back into play here, so I will do the same: The nodes are the two points of intersection between a planet's orbital plane and the ecliptic plane. "Every digression in latitude is measured from the nodes";

therefore the ascending node is the point when the planet "enters into northern latitudes," and the descending node is the opposite. In short, by "digressions," Copernicus means "digressions from the ecliptic." For an illustration, consider two digressions of Mars which Ptolemy observed: In solar opposition "at the farthest limits of southern latitude," the figure was 7°; in solar conjunction it became a mere 5', "so that it almost touched the ecliptic." (By the way, this planet shows the greatest digression of all.)

The latitude "which occurs at the mean longitudes" Ptolemy called the **orbital inclination** (we now define it as a celestial orb's angle of tilt with the ecliptic); the latitude at the highest and lowest **apsides**, which are the points in an orbit when two celestial orbs draw closest together, he named the **obliquation**, which Jacobsen more helpfully explains as "a small periodic fluctuation in the inclinations of the planetary deferents," "and the third one, which occurs in conjunction with the second," Ptolemy termed the **deviation**. Jacobsen again: "Fluctuation of the planes of an epicycle." In our chapter on the orbits of Venus, this astronomer has already savaged deviation for us. Indeed, neither obliquation nor deviation finds much employment these days. (For this reason and others, Book VI is a thoroughly ghastly book.) Not foreseeing the future, Copernicus adopts these terms to explain the regular movements of the planets' orbital circles around the ecliptic. In the case of the three outer planets, libration comes into play—always a benefit to any devotee of uniform circular motion, for then he can say: "In the case of things which are undergoing a libration, we must take a certain mean between the extremes." For their part, Venus and Mercury display another sort of libration. Needless to say, the discussion now becomes so simultaneously technical and specific as to rise beyond my meager sublunary scope; each planet gets singled out in terms such as: "Mercury also differs from Venus in that its libration takes place not in a circle homocentric with an eccentric circle but in a circle eccentric to an eccentric circle." But in all cases, libration gets very logically defined as difference between maximum and minimum inclination; half of this figure equals the mean inclination.

Copernicus now proceeds to use Ptolemy's observations of digression in order to calculate the inclinations of the orbital circles of Saturn, Jupiter and Mars, from which he'll determine angles of apparent latitude:

Planet	Max. Inclin.	Min. Inclin.	Modern Val.
Saturn	2° 44'	2° 16'	2° 29'
Jupiter	1° 42'	1° 18'	1° 19'
Mars	1° 51'	9'	1° 26'
Venus	3° 29'	46'	3° 12'
Mercury	6° 15'	7°	7°

The modern inclination values obviously correspond fairly well to Copernicus's maximum inclination values.

As for deviation, given that that is now nothing but the decomposed cadaver of an erroneous concept, I will content myself by regaling you with the following eternal maxim: "The deviations of Venus always remain northern and those of Mercury southern." While the observations from which this sprang remain true, the concept itself was on par with Ptolemy's equant.

VI.9: "Except that in the case of Mercury ..."

There follow two pages of generalized instructions "on the calculation of the latitudes of the five wandering stars." As we know, when Copernicus says "latitude," he might be referring to the latitude of declination, of obliquation or of deviation. But now he means the "predominant latitude." Calculate the other three and add them if they are all positive or negative; if not, subtract the different one from the two of the same sign, "and the remainder will be the predominant latitude sought for." That is how *Revolutions* ends—typically, exasperatingly Copernican! The tables for the latitudes of Saturn, Jupiter and Mars share some column labels in common with the corresponding tables for Venus and Mercury, but the latter have deviations and obliquations whereas the former have northern and southern subdivisions. The outer planets have "Proportional Minutes"; the inner boast "Proportional Minutes of the Deviation." The instructions themselves are rife with such mind-bogglers as ". . . except that in the case of Mercury one tenth of the obliquation is to be subtracted, if the anomaly of the eccentric circle and its number are found in the first column of the table, or merely added, if in the second column of the table; and the remainder or sum is to be kept." In short, "the mind shudders."

Simplicity

Suppose that our Earth did move instead of the sun? "As far as the appearances of the stars are concerned, nothing would perhaps keep things with being in accord with this simpler conjecture," Ptolemy readily concedes. All the same, "in the light of what happens around us in the air such a notion would seem altogether absurd." Recall that Occam's Razor (which once again I admit was formulated long after Ptolemy died, and which once again I assert that Ptolemy and his successors generally tried to follow) advises us to accept the simplest hypothesis which conforms to the facts as we understand them. Or, as Kepler so practically put it, "Astronomy has two ends, to save the appearances and to contemplate the true form of the edifice of the world." The limits of observation in A.D. 151 had not yet been pushed back sufficiently to disprove Aristotelian postulates concerning the motions of "what happens around us in the air." Therefore, Ptolemy lent his impressive intellect to "saving the appearances" of a plausible but erroneous interpretation of the universe. Erroneous, yes, but plausible indeed. An astronomer from our own time remarks that epicycles saved the geocentric system for two thousand years by agreeing "fairly well with the angular motions of the planets in longitude, with an accuracy about equal to that of contemporary observations . . . In fact, they may be said to have been scientifically acceptable when judged by the modern criterion of satisfactory agreement with observations." All the same, the appearances were not quite saved. As we have seen, simplicities accordingly had to get less simple over time, as we added more epicycles and equants to explain whatever inconsistencies came to light.

Hence *Revolutions*.

Astrologers' shameful recourse

What happened to Copernicus's crazy notions after they had rotted to perfection in their coffer? "For a time," says Kuhn, who means to tell the tale of all scientific revolutions, "they were used by the specialist even though, within the larger climate of scientific thought, they seemed incredible." Indeed, less than a decade after *Revolutions* was printed we find Erasmus Reinhold's *Tabulae Prutenicae Coelestium Motuum* employing Copernican geometries to compute superior planetary tables, all the while agreeing with the rest of us that of course the Earth stands still. "In every work there are to be observed the situation, motion, and aspect of the stars and planets, in signs and degrees, and how all these stand in reference to the length and latitude of the climate; for by this are varied the qualities of the angles, which the rays of the celestial bodies upon the figure of the thing describe, according to which celestial virtues are infused." These words were written by an occultist a full 258 years after the publication of *Revolutions*. They make clear what astrologers require, and why. And what they wanted, *Revolutions* in significant measure gave them. Since the sun has its greatest power over us as it passes into the nineteenth degree of Aries, while it is feeblest in the nineteenth degree of Libra, Copernicanism's slightly superior accuracy can decrease our likelihood of planning a wedding for a Thursday morning when we should have chosen a Wednesday afternoon. Copernicus denies the sun's movement, you say? No matter. We employ the tool, not the toolmaker. (Does the sun move? Our occultist just quoted—you'd think that in 1801 he'd feel beaten down—remains defiantly evasive.)

By about 1575, the so-called "Wittenberg interpretation" of Copernicus, invented by such luminaries of Wittenberg University as Melanchthon, submits to the mathematics of Books II through VI while continuing to defy the heliocentrism of Book I. When the Gregorian calendar comes into use in 1582, the alteration of the year is partially indebted to Copernican mathematics. Indeed, Copernicus's acolytes will go so far as to say that it is thanks to his discovery of the length of the tropical year that the Gregorian calendar becomes accurate to one day in three thousand years. On the other side of posterity's courtroom, Jacobsen informs us that "no considerable improvement followed immediately from the adoption of the heliocentric viewpoint . . . mainly because the distances and longitudes, still based on circular deferents without equants, were prohibitively in error." As the seventeenth century dawns, this deficiency heartens those who prefer to interpret Copernicus's sun-centered follies in Osiander's spirit.

Jacques Barzun believes that *Revolutions* "proposed an important change indeed, but it was not the shattering blow it is commonly taken for; it raised new difficulties, and those who rejected it were not simply diehards refusing evidence." Bit by bit, however, the cloistered conclave of Copernicans came to include the active, the visible and the worldly; for the new difficulties were met one by one.

By the time Copernicus wrote *Revolutions*, the map of the Ptolemaic cosmos bore as many patches and plasters as Ptolemy's corpse would have had to wear if some sixteenth-century lover of perfection had tried to make it walk. *Revolutions*'s raison d'être was simplification.

"Except that in the case of Mercury one tenth of the obliquation is to be subtracted . . ." All the same, *Revolutions* succeeded insofar as it did simplify, and ruthlessly. It failed insofar as it clung to such verities as eternal circularity, which "the appearances" might not exactly disprove (such being the limits of observation before Tycho Brahe), but which they certainly didn't prove, either. "It will be necessary for us to assume irregular and corrected movements everywhere as the differences in velocity and to employ them in the demonstrations," Copernicus admits.

"His planetary theory could not reproduce the observed values any more exactly than the older theories," complains a twentieth-century encyclopedia of astronomy. But even the astrologers, to whom an uncentered Earth is anathema, embrace his system, such is its predictive power. Before we know it, Copernicanism is out in the open! By 1619, the great cartographer Blaeu represents the Copernican system in a map of our home planet . . .

Epilogue to Mercury's obliquation

Were we to describe the so-called "Copernican Revolution" in brief, we might put it this way: That predictive power grew ever more irresistible. Ptolemaic astronomy for its part vainly altered this or that predication, its thrashings merely reactive. Kepler's excellent approximation of the planetary orbits—elliptical paths through which each celestial body passed through

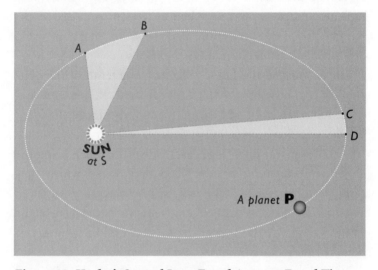

Figure 19   Kepler's Second Law: Equal Areas at Equal Times
The orbits of the planets have finally become elliptical! Can any scrap of uniform circular motion be saved?

   Kepler admits that P speeds and slows as it whirls around the center of the sun, S. However, angle ASB cuts off the same area as the longer, narrower angle CSD. Therefore, P must traverse the arc AB in the same amount of time as when it traverses arc CD. P's velocity will thus be greater over AB than over CD. This phenomenon will soon have a fundamental explanation: gravity.

equal areas at equal times—combined with Copernicus's helio-centric assumptions to explain the perceived and real variations in planetary velocity so well that epicycles and equants were no longer needed at all. The principle of Occam's Razor had operated once more; why draw circles circling circles if we could get by without them?

Once Newton had compressed Kepler's principle into the still more economical and universal inverse-square law (gravitational attraction falls off as the inverse square of the distance), it became possible to show why Kepler's approximation was but an approximation: Each ellipse was distorted by the gravitational pull of neighboring planets! Studying these distortions led to what Kuhn calls "one of astronomy's greatest triumphs": In 1846, Neptune was found through pure predictive mathematics. Ptolemy's system could never have done that.

Oh, but it all happened so slowly! When Sir William Herschel (the father of the Herschel whom I so often mention in this book) discovered Uranus in 1781, that was by accident . . .

Back to iron-grubbing

Let's be cheerful, and say that it happened not only thanks to intellectual inertia, both of motion and of rest, but also because there is no such thing as a dead end in science—or, as I should say, dead ends convey information. We live and think in a labyrinth. Every time one of our turnings reaches a blank wall, our knowledge of the maze increases. Astrology, alchemy, Lysenkoist biology and the Ptolemaic spheres all taught us something. Most of the works of Ptolemy, Copernicus and Kepler to which I've referred in this book are all bound together in one volume, part of the Great Books of the Western World series; and in a very real sense they *are* all one volume. In a certain footnote to a discussion of epicycles and equants in *The Almagest*, the translator remarks: "These three points—the center of the equant, the center of the deferent, and the center of the ecliptic—with the deferent's center midway between the other two, by means of the Copernican transformation, come to represent the two foci and the center of the Keplerian ellipse with exactly the same ratios preserved." The fact that the ratios are preserved stuns me. How then can we say that Ptolemy was false? What happened is not that Copernicus threw Ptolemy in the rubbish-heap, then Kepler likewise disposed of Newton, but something else altogether, something strange, beautiful, perhaps even spiritual.

At the beginning of this book I suggested that browsing through the past's treasure-heap could do us no harm. You and I, reader, have turned over quite a lot of grimy detritus together. I hope that you have come to believe, as I do, that everything does retain some value, no matter how rusted over with discredit it may be.

For instance, in the words of Sir John Ball, "The great difficulties which beset the monstrous conception of the celestial sphere vanished, for the stars need no longer be regarded as situated at equal distances from the Earth." But we have seen that mariners continue to use the celestial sphere to this day, because its assumptions simplify navigation. An astronomer writes me that even he employs the celestial sphere on occasion, "because we can measure an object's position in two dimensions (i.e. where it is projected on the sphere) extremely well, while our distance measurements are still extremely crude by comparison." Meanwhile, Kepler retains the Sphere of the Fixed Stars, considering it "a kind of skin of the world," or again "the riverbed in which" the sun's "river of light runs," even though he finally rejects the notion that the spheres have any corporeality, since Tycho's observation of comets proved that those bodies cross the spheres' supposed boundaries; furthermore, those vexed backward wriggles of Mars's orbit require the Sphere of Mars to intersect the Sphere of the Sun. So why does he need the Sphere of Fixed Stars? It may well be because he is a Scriptural astronomer. He needs his own species of simplicity.

What about geocentrism? Even that survives inasmuch as "the Earth may be a Hawaii in a universe of Siberias." That simplicity bears its own human truth, whether or not it turns out to be scientifically valid. If we can cherish our planet for its very uncentered vulnerability, can't we get back some of what we lost after we stopped believing in geocentrism literally?

And why does Galileo continue to insist on orbits of perfect circularity? What about all the false circles at the heart of this book? Aren't they rubbish? Herschel calls Copernicus's theory "rather a geometrical conception than a physical theory, inasmuch as it simply assumes the requisite motions." No gravity; I've said that a dozen times; then this absurdity called uniform circular motion—there *is* no uniform momentum for the revolutions of the heavenly orbs. But the case is different for **angular momentum**, which may be defined in relation to the center A (in this case, the sun) about which the planet moves, as

$$\vec{J}_a = \vec{p}\vec{D},$$

where $\vec{J}_a$ is a vector force (which travels perpendicular to the plane of the sun), $\vec{p}$ is a momentum vector (and momentum breaks down into the mass of the planet times its orbital speed), while $\vec{D}$ is the linear distance from the sun through the planet in the direction of $\vec{p}$.

And angular momentum is conserved whenever only one force acts upon the heavenly orb in question, and when that force remains aimed at the sun. Kepler's "equal areas at equal times" simply means that angular momentum about the sun is uniform.

In a sense, what Copernicus in his dedicatory letter to the Pope had referred to as "the holy principle of uniformity of motion" remains alive; only it has been first broadened and abstracted by Kepler, then universalized by Newton into something still more true to the appearances, if less apparent to the false simplicity of our senses, which insist that the universe revolves around us.

But the universe screamed

So much for the long view, which is its own distortion.

Please consider the book thus far an introduction. We now have sufficient information to diagram, however crudely and intuitively, the vector forces acting through *Revolutions*. In the next and final chapter, we will tell the story, not of Copernicus or the Copernican Revolution, but of Copernicanism.

Burnings

... the boundary posts of true speculation are the
same as those of the fabric of the world; but the
Christian religion has put up some fences around
false speculation ... in order that error may not rush
headlong.

—*Kepler* (1618–21)

The Medicean planets

At "the first hour of the night," in the first month of the auspicious year 1610, Galileo "perceived (as I had not before, on account of the weakness of my previous instrument) that beside the planet" Jupiter "there were three small starlets, small indeed, but very bright. Though I believed them to be among the host of fixed stars, they aroused my curiosity somewhat by appearing to lie in an exact straight line parallel to the ecliptic, and by their being more splendid than others of their size." Over the course of the next few nights, these stars changed in position and in number, but never drew far from their host, "accompanying that planet in both its retrograde and direct movements in a constant manner," which is why "no one can doubt that they complete their revolutions about Jupiter and at the same time effect all together a twelve-year period about the center of the universe."

Galileo seizes on these four "Medicean planets" of his (fifty-nine more had been discovered by the time I wrote this book, twenty-three of which came to our attention in 2003) to rebut the critics of Copernicus who "are mightily disturbed to have the moon alone revolve around the Earth . . . Some have believed that this structure of the universe should be rejected as impossible."

We upholders of a centered Earth might comfort ourselves with the notion that at least the local orbits of these novelties substantiate Ptolemy's—and Copernicus's—belief in epicycles, but even if traditional geometric conveniences can be saved, the fact remains that the revolutions of heavenly bodies need not be concentric! And so the discovery of Jupiter's moons becomes the next wound in our ancient, perfect universe, whose spheres once revolved attentively about us and our destiny.

Until now, Ptolemy's conclusion of a "sensibly spherical," motionless Earth in the geometrical center of the heavens, which "are spherical and move spherically" appeared to reconcile theory quite well with common sense. *Revolutions*, not only less elegant but also less clear than *The Almagest*, never had common sense on its side. But how can Ptolemy defend these decidedly non-geocentric movements of Jupiter's moons? The limits of observation, you see, have just receded.

Once upon a time, Ptolemy wrote: "Now, we have used the things previously demonstrated concerning the sun as if it displayed no sensible parallax—not because we were unaware that its parallax when later worked out would make some difference in these matters, but because we did not think any appreciable error would follow from this with respect to the appearances." Let his sentiment be taken as the beginning of a parable. And now that future time when everything begins to get worked out is upon us, and innocuous little errors turn out to be appreciable after all. That's the parable's end. Who would have thought it? The same fate will overtake Copernicus.

How timorously we uncenter ourselves! Copernicus keeps the faith that the Earth remains so fundamentally important to our neighboring planets that terrestrial mobility "binds together the order and magnitude of their orbital circles in a wonderful harmony and sure commensurability." (How much credit do you think we should give our puny planet for the shape of Jupiter's orbit?) And remember Tycho, whose years of celestial observations cannot dissuade him from recentering the Earth in the Copernican cosmos. Why not? Copernicus himself couldn't go farther than this: "Let the center of the world"—which means, remember, the center of the *universe*— "be F," and F is almost at the sun. In 1621, the great Kepler will only be able to leave geocentrism sufficiently far behind to "teach the conformation of the whole universe" with "the place of the sun at its center," a position he defends with reference to the analogies of Scriptural astronomy: The sun, the stars and the planets in the space between them equal the Father, the Son and the Holy Ghost! Now astronomy teaches that our sun is an anonymous speck of gas.

We think of the universe as a vast, dark place, intermittently speckled with stars and dust. Copernicus's "world," like Kepler's, still remained centered (at or near the sun), intimate and alive with sunlight: "For since the other parts of the world are pure and are filled with the light of day, it stands to reason that night is nothing except the shadow of the Earth, which has the figure of the cone and ends in a point."

Once upon a time, this universe was whole; it was *ours*. We know from Scriptures that God "gives the sun for light by day and the fixed order of the moon and the stars for light at night." By the power of that Word, and by the logic of *The Almagest*, the Sphere of the Moon would never alter until Judgment. But in this unhappy era I live and write in, specialists whose procedures we lack the education to judge announce that our moon is gradually drawing away from us! There is no fixed order of anything. And that's why Luther cries: "That fool wants to overthrow all the art of Astronomy! But as the Holy Writ shows, Joshua ordered the sun to stop and not the Earth."

Poor Copernicus! How the Protestants despise him! Philipp Melanchthon says: "Prudent sovereigns should bridle such licentiousness of the human mind." Calvin is said to have denounced him thus: "Who will venture to place the authority of Copernicus above that of the Holy Spirit?" (But I cannot get hold of Calvin's works in the Varmia-like backwater where I dwell; and another authority assures me: "Never having heard of him, Calvin had no attitude toward Copernicus.")

Once upon a time, the length of our days and nights might vary by season, but the cycle always repeated itself. Now those same specialists inform us that this cycle was deficient in static perfection and had indeed been altering long before the first man and woman fell into sublunary sin: Earth's rotational period used to be twenty-one hours in the Cambrian Era. Someday it will be sixty days.

Everything is sublunary now. Everything is falling away from us and decaying. And we take it as a matter of course. Shot through with sublunary spirit, an astronomy textbook from the late twentieth century concludes: "A philosophy that sees mankind as a permanent fixture in an unchanging environment, while it may be practical for short periods such as lifetimes or centuries, cannot be defended on a cosmic scale." What would the churchmen who already hated Copernicus have made of that?

What undermined religious faith? asks the philosopher Emil L. Fackenheim. He answers his own question as follows: "Most people would say: modern science. The story begins with Copernicus, who shows that the Earth is but one of many stars [*sic*]; it is carried forward by Darwin . . . and it culminates with Freud . . ."

Jacques Barzun for his part opines that Copernicus did us a favor by removing us from the center of the universe, "when men thought of themselves as miserable sinners fearful of an angry God." Still more optimistically, an early-twentieth-century physicist sees the destination of the Copernican Revolution as "the ultimate triumph of celestial mechanics— a triumph which robbed the planets of their age-long dominance over the lives of man." But even if these men are correct, and the attentions of the old cosmos were stern, never mind malign, at least they were attentions. Now we began to fear that Augustine's "firmament of authority over us" might be empty. What if Father has gone away? Somewhat more than two centuries after Galileo discovered Jupiter's moons, Nietzsche will write: "God is dead."

What if space were mainly emptiness, and even the atoms we're made of mainly empty space? We couldn't bear that. (I quote the *Twentieth Century Encyclopedia of Catholicism*: "We now know that the sun, the star of which the Earth and planets are satellites, forms part of an enormous system of stars known as the Galaxy.") Against this possibility, Kepler desperately asserts: "The sun is the principal body of the whole world."

Resolutely Copernican

The universe screams. But it will defend itself, not least through the new Pope, Urban VIII, and his deputies, whom a historian more charitable than I sees as "the first bewildered victims of the scientific age." First, last and most predictably, it denies, rejects the wound. At Leiden, Pieter de Bert publishes a book in 1604 which places an unmoving Earth at the center of the universe. We're informed that only Cracow, Oxford and Salamanca never openly stood against the Copernican system. And it may be that even this list is too long, for in 1583, and then again the following year, when a contentious reader of forbidden books named Giordano Bruno (proof that he's an imminent danger to our centered universe: he's already been excommunicated not only by the Catholics but also by the Calvinists) dares to lecture on Copernicanism in Oxford, he and his hearers quickly start shouting at one another. Oh, yes, the universe stands fast; the Earth remains unmoved at Oxford!

Young Tycho Brahe has long since discovered that Copernican predictions outshine the accuracy of the old Alphonsine Tables. As we've seen, he lacks the means to measure the stellar parallaxes which Copernicus's theory requires, so, clinging to the wrecked hulk of the old universe, he proposes that while some planets might revolve around the sun, the sun and everything else continue to circle the Earth.

To us, the Tychonic system appears as artificial as Tycho's nose, which is comprised out of electrum to replace the organ nipped off by a duelist's sword; all the same, it does as good a

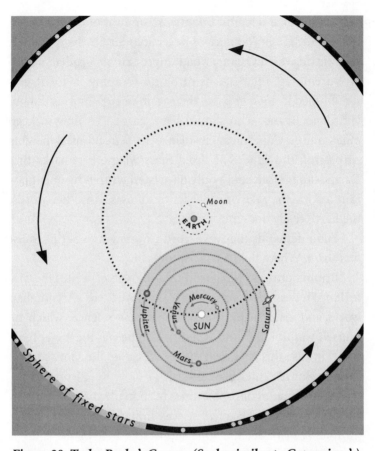

Figure 20 Tycho Brahe's Cosmos (Scale: similar to Copernicus's)
Uniform circular motion still assumed. Epicycles, eccentricity,
etcetera—not shown.

job as Copernicus's of "saving the appearances." No doubt a
number of us are grateful for the limits of observation which
make Tycho's system plausible for several years longer. A few
pietists will keep the centered faith all the way into the nine-
teeth century.

The universe screams, but attacks upon it only grow more brazen. Although *Revolutions* never sold out its first printing, the one on which Osiander had impressed his prudent stamp at Nuremberg, there seems to be no escaping it, for it gets republished in 1560 in Basel, then again in 1617 in Amsterdam. Giordano Bruno, who should have learned by now to keep quiet, raises Copernican insolence to astronomical heights surpassing the Sphere of Fixed Stars when he remarks that Osiander's preface could only have been written by one ignorant ass for the sake of other ignorant asses. In other words, the Earth really does move!

Bruno demands that we accord Copernicus a Scripturalist literalism. Where then will Scriptures be?

"Bruno's attitude was as radical as possible," writes the historian Dugas. "He completely destroyed the Aristotelian world and cleared the ground for a new science, which he himself was not to found, but of which he was the metaphysician." In case this isn't clear enough, I quote the same source: "Bruno was resolutely Copernican."

The distinction has sometimes been made that Copernicus continued to say the universe was finite; Bruno, that it was infinite. No wonder that "the suppressed rage of the Church was turned on Bruno."

In 1589 he gets excommunicated for the third time, this time by the good Lutherans of Helmstedt; in 1590 he's refused permission to reside in Frankfurt; in 1591 a patrician lures him back to Italy and turns against him. In 1592 his trial for heresy begins in Venice. The Inquisition extradites him to Rome, where the charges against him are meticulously constructed—see, we saviors of centeredness are as good as scientists!—over the next seven years. The Pope condemns him. On 8 February

1600 he famously tells his Inquisitors: "Perhaps your fear in passing judgment upon me is greater than mine in receiving it." Nine days later they burn him at the stake, with a gag upon his tongue so that his final cries can wreak no further harm upon the universe.

"How great would have been thy joy"

Missing the warning implicit in Osiander's preface, Galileo rushes on in the same literalist direction as Bruno: "Oh, Nicolas Copernicus, how great would have been thy joy to see this part of thy system confirmed with so manifest experiences!" How great would have been Copernicus's *terror*, he should have said. How great his own should have been . . .

He is known to have contacted that other dangerous subversive, Kepler, to whom he confesses to be a Copernican covert. No, he hasn't missed any warning; in a letter to Kepler he admits his fear . . .

And his telescope exposes the moon's astroblemes, the sun's spots. It's been said that "Copernicus finds his main argument for his rearrangement of the cosmos in its presumed harmony." But look where that harmony has led us! Sublunary leprosy is everywhere. No wonder so many refuse to peer through Galileo's lens. I don't suppose that Foucault's pendulum would have been welcome, either.

"Newly emerging values still seeking intellectual justification"

In 1593, Kepler writes a dissertation about the moon and the revolving Earth. A fellow student requests permission from the faculty and administration at Tübingen to hold a debate on the subject. Upholding the universe, this request the good Lutherans deny. Not long afterward, they suppress the first chapter of Kepler's first book, *Cosmic Mystery*, because it argues against anti-Copernican interpretations of the Scriptures.

And so another career continues as expected. In 1605 Kepler finally accepts necessity and postulates elliptical orbits, but, just as Copernicus would have done, delays publication. In 1611, he finds himself rejected for a professorship at Tübingen, because he might well "arouse much unrest at the University." Or, as has been more blandly put (the subject was Copernicus's predecessor, Nicholas of Cusa), "Traditional and conservative elements from the medieval era not yet ended meet, often clash, with newly emerging values still seeking intellectual justification."

How many ways and how many times must I make this point before it becomes emotionally explicable, not only to you, my unknown reader, but also to myself? How can we hope to feel this other universe which Copernicus helped to destroy? We never lived in it; we can hardly imagine it. "Man is the measure of all things." What physicist, what chemist or biologist practices his specialty on those terms today? A historian of early Christian symbolism relates that once upon a time, "the entire cosmos . . . consisted of signs of God or could become a divine sign." We remain free to imagine what the Pseudo-Dionysius the Aeropagite envisioned a thousand years before Copernicus: the lightning-flash and the flame, the ineffable, the unknown. These do exist, in supernovas and black holes, in the unexplored vastness of space. To me they are enough. To the learned Doctors of Tübingen, how could they have been sufficient? Where was their Earth-centered symmetry?

In 1619, Kepler's *Epitome of Copernican Astronomy* is entered onto the Index of banned books. This is a man whose mother was tried for witchcraft and even brought into the torture chamber. Oh, he's uncanny, all right; he's dangerous! "My books are all Copernican," he confesses in a private letter.

"Safely back on a solid Earth"

This little book of mine began with the parable of Osiander's preface. It now ends with it.

We might think of "Scriptural astronomy" and heliocentrism as two celestial bodies circling round and round the common sun of the Unknown: Sometimes they reach a point of inferior conjunction, and Copernicus was lucky enough to do his work at one of those times; but their differing orbital motions inevitably draw them ever further apart, until at last they come into superior conjunction; Bruno and Galileo met their separate dooms because of this accident of time and place just as much as for their fanaticism.

In 1536, Cardinal Nicholas Schönbeg may have benignly offered to subsidize the publication of Copernicus's writings; whereas in 1612, Paolo Gualdo warns Galileo: "As to this matter of the Earth turning round, I have found hitherto no philosopher or astrologer who is willing to subscribe to the opinions of Your Honor, and much less a theologian; be pleased therefore to consider carefully before you consider this opinion assertively, for many things can be uttered by way of disputation which it is not wise to affirm."

In other words, Osiander's preface, like the moon and stars, was placed over us for a reason!

But Galileo continues to affirm that which it is not wise

to affirm; Copernicus for his part had declined Cardinal Schönbeg's kindness. What opposite stories those two heroes tell!

One modern astronomer opines with eminent justification that Copernicus "was comfortable with the idea of publishing tabular values of planetary positions but much less so with the prospect of provoking his colleagues by publication of new theories." Some say that Copernicus feared nothing more than their scorn; others, that "there should be no illusions about the temper of those times . . . Copernicus was aware of the dangers which surrounded him." Oh, yes; he catwalked ever so gingerly through the disputed zone! In 1539, the sympathetic Rheticus prepared a resume of the first three-quarters of *Revolutions*, referring to him, one must assume at his instigation, only as "my teacher" or "the Herr Doktor."

For his part, Galileo not only seems to feel no such fear, he waxes protectively *indignant* on behalf of the dead Copernicus, whom he calls "our master." Many things can be uttered by way of disputation which it is not wise to affirm; Osiander saved Copernicus from affirming them, but Paolo Gualdo owns no such power over Galileo, who in a prefiguration of his own epitaph remarks that Copernicus "procured for himself immortal fame among a few but stepped down among a great crowd (for this is how foolish people are numbered), only to be derided and dishonored."

Thus Copernicus; thus Galileo, who presses on, preparing to deal our universe new wounds!

Not only does he discover sunspots; worse yet, he tracks them, which proves the sun's rotation about its own axis. Well, that still doesn't kill centeredness; Ptolemy gave permission for celestial orbs to twirl while they spun about us . . .

Galileo complains about those who, "replete with the pertinacity of the asp," refuse to look through his telescope. Doesn't see that their resistance to seeing could yet be his preservation, as it was Copernicus's?

Cardinal Bellarmino sends a friendly note to Paolo Antonio Foscarini: "It seems to me that your Reverence and Signor Galileo act prudently when you content yourselves with speaking hypothetically and not absolutely, as I have always understood Copernicus spoke."

In other words, we could have lived with heliocentrism if it had only hid its nakedness a trifle, like a Polish church-tower rising half-seen behind mist and trees.

Still Galileo will not listen.

The Inquisition, with its disregard for the rights of the accused and its proclivity to swallow victims on mere suspicions, is in one historian's words "a system which might well seem the invention of demons." Regarding Galileo's case it must be admitted that he repeatedly convicted himself, untiringly committing the sin of speaking absolutely and not hypothetically. And so they rescue us from him.

After analyzing two volumes of sentences by Roman and provincial Inquisition tribunals for the years 1580–82, the scholar Tedeschi concludes that almost half of these judgments were rendered in the Venetian Republic, almost half for Protestantism, slightly over a quarter of them for magic and witchcraft, and almost none (10 out of 225) for "opposition." In short, just as we would expect with regard to sublunary matters, localism reigns, and the most punitive attention gets devoted not to the black arts and suchlike forms of extreme subversion, but to a rival version of the same creed. Now, what about that minority category of opposition? This would seem to be Galileo's sin. Was defiance of authority really so rare in those days?

Oh, he's a rare one, all right. "Our problem is to find arc *FC* of half the retrogradation," murmured Copernicus to other savants, but Galileo shouts: *The Earth moves!*

And so they menace him into abjuring truth:

I have been pronounced by the Holy Office to be vehemently suspected of heresy, that is to say, of having held and believed that the Sun is the center of the world and immovable and that the Earth is not the center and moves . . .

—an evil doctrine, a Copernican one, which he kneelingly renounces. His future: disgrace, fear, house arrest until death. The universe has been saved.

A certain Querengo complacently writes in 1616, once he hears that *Revolutions* has been banned "until corrected" (Kepler's *Epitome* will be likewise proscribed two years later): "So here we are at last, safely back on a solid Earth, and we do not have to fly with it as so many ants crawling around a balloon . . ."

Chronology

ca. 120 B.C.	Hipparchus discovers precession of the equinoxes.
384 B.C.	Birth of Aristotle.
347 B.C.	Death of Plato.
ca. 340–320	Aristotle writes the *Physics, On the Heavens* and other treatises.
322 B.C.	Death of Aristotle.
ca. 151	Ptolemy finishes *The Almagest.*
ca. 1200	Birth of Albertus Magnus.
1252–62	King Alphonso X of Castile sponsors Jewish and Arabic astronomers in their calculations of the Alphonsine Tables, which predict planetary positions.
1415	Execution of Jan Hus for heresy.
1416	Execution of Jerome of Prague for heresy.
1473	**Copernicus born in Torun.**
1483	Copernicus's father dies.

1483	The first printed version of the Alphonsine Tables (published in Venice) comes into the world.
1491?	Copernicus begins to study liberal arts at the Jagiellonian University in Cracow.
1492	Columbus discovers America.
1494	Copernicus studies law at the University of Bologna.
ca. 1512	Copernicus commits his heliocentric idea to writing in the *Commentariolus*.
1522	Circumnavigation of the globe (begun by Magellan expedition).
1540	Publication in Gdansk of Rheticus's *Narratio Primo* (written the previous year), which summarizes parts of *Revolutions*.
1542	Pope Paul II "reinvigorates" the Inquisition.
1543	**Publication of Copernicus's *Revolutions of the Celestial Spheres*, almost simultaneously with the death of Copernicus.**
1545–63	Council of Trent requires Catholics to adhere both to Scriptures and to unwritten Church traditions.
1546	Tycho Brahe born.
1551	Erasmus Reinhold's *Tabulae Prutenicae Coelestium Motuum* uses Copernican mathematics while denying heliocentrism.
1553	Michael Servetus burned at the stake for heresy involving, among other things, astronomy.
1564	Galileo Galilei born.
1571	Johannes Kepler born.
1577	Tycho asserts that a heliocentric solar system revolves around the Earth.
1600	Giordano Bruno burned at the stake.

1601	Tycho Brahe dies.
1610	Galileo invents the telescope.
1613	Galileo's "Letters on the Solar Spots" prove the rotation of the sun.
1615	Inquisition begins first investigation of Galileo.
1616	**The Qualifiers of the Holy Office in the Vatican declare heliocentricism and a moving Earth to be absurd and heretical. Copernicus's *Revolutions*, which happens to be infested with such errors, is decreed to be "suspended until they can be corrected" and placed on the Index of banned books.**
1618–21	Publication of Kepler's *Epitome of Copernican Astronomy*.
1620	***Revolutions* is "corrected" by excision of nine assertions that the book is literal fact.**
1630	Kepler dies.
1633	Galileo recants his Copernican heresy.
1639	First observation of a transit of Venus across the sun.
1642	Galileo dies.
1642	Isaac Newton born.
1727	Newton dies.
1757	Pope Benedict XIV repeals prohibition of Copernican books but not of Copernicus's uncorrected treatise.
1781	Sir William Herschel discovers Uranus.
1835	**Copernicus's book pulled off Index.**
1838	F. W. Bessel measures the parallax of the star 61 Cygni, thereby proving the annual motion of Earth.
1846	Discovery of Neptune.

1849	Sir John Herschel writes his *Outlines of Astronomy.*
1851	The first empirical proof of the Earth's diurnal rotation is accomplished with Foucault's pendulum in 1851.
1915	A. H. Compton proves the Earth's diurnal rotation in a different way, this time by means of a rotating water-filled tube whose contents display the Coriolis Effect.
1930	Discovery of Pluto.
1934 or 1935	Making of first photograph (infrared, from a balloon) which depicts the curvature of the Earth.

Glossary

The provenance "Ptolemaic" is shorthand for something which was incorporated into the Ptolemaic system. Sometimes it may have been invented by one of Ptolemy's predecessors (e.g., the concept of the eccentric). The same goes for a "Copernican" item.

Albedo The proportion of light which gets scattered from the surface of a planet or star.

Angular elongation The angle along the ecliptic between a planet and the sun, or a planet and a satellite, as measured from the Earth (in degrees east or west of the sun).

Anomaly [Ptolemaic and Copernican]. A regular motion which in combination with another regular movement causes the latter to seem irregular.

Aphelion The point in an orbital path when a planet, moon, asteroid, etcetera is farthest from its sun.

Apogee The point in an Earth-centered orbit when the celestial orb is farthest from the Earth.

Apsides [Ptolemaic and Copernican]. The points in an orbit when two celestial orbs draw closest together and when they grow farthest apart.

Astrobleme   A "blemish" on the surface of a celestial body, caused by the crash of another body such as an asteroid.

Astronomical unit   A measurement based on the ecliptic radius. The Earth's mean distance from the sun is 1 AU.

Circle, great   The intersection of a sphere and a plane which passes through the center of a sphere; hence the "equator" which bisects the sphere into two symmetrical hemispheres.

Conjunction, inferior   When two planets are closest (in a straight line on the same side of the sun), the inner one is at inferior conjunction to the outer. The outer one is in opposition to the inner. When Earth is one of the two planets, the conjunction is said to be with respect to the sun.

Conjunction, superior   When two planets are farthest (in a straight line on opposite sides of the sun from one another), the inner one is at superior conjunction to the outer. The outer is at conjunction to the inner. When Earth is one of the two planets, the opposition is said to be with respect to the sun.

Declination   A celestial body's angle with respect to the plane of the Earth's equator. A projection of terrestrial latitude on the celestial sphere.

Deferent, circular [Ptolemaic].   The great circle of a celestial body's orbit which bears one or more **epicycles.**

Deviation [Copernican].   Fluctuation of the planes of an epicycle. Copernicus refers to it as the third type of latitude, which occurs in conjunction with the second, which is **obliquation** (VI.1).

Eccentric [Ptolemaic and Copernican].   A **deferent** which is not centered on the Earth, sun, or other celestial orb about which the deferent revolves.

Eccentricity, orbital   "The deviation of an orbit from a circle as given by $(L - S)/(L + S)$ where L and S are the long and short diameters of the orbit."

Ecliptic [Ptolemaic *vs.* Copernican]. The plane of the Earth's orbit about the sun. Copernicus defines it as "the circle through the middle of the signs under which the center of the Earth moves in a circle in its annual rotation." Ptolemy of course defines it in the opposite way, as the sun's apparent course around the Earth.

Ecliptic, oblique. *See* **obliquity of the ecliptic**.

Epicycle, planetary [Ptolemaic]. A circular sub-movement postulated to make an irregular orbit correspond to perfect circularity. Any number of epicycles could be added to "save the appearances." Think of wheels within wheels. An example would be the orbit of the moon about the orbiting Earth.

Equant [Ptolemaic]. A mathematical point, distinct from the center of a planet's circle of rotation, about which the planet's motion is uniform. This rather strained method of "saving the appearances" was rejected by Copernicus.

Equator, celestial. See **sphere, celestial**.

Equinoxes, vernal and autumnal The two points of intersection between the ecliptic and the celestial equator; these Ptolemy defines as "the equinoxes of which the one guarding the northern approach is called spring, and the opposite one autumn." At those two antipodal points where the ecliptic corresponds to the celestial equator, which according to our current calendar occurs on 21 March and 23 September, there, and only there, day and night are equal in length all over the Earth. (Day and night are always equal at the terrestrial equator.)

Inclination, orbital [Ptolemaic and Copernican]. The latitude "which occurs at the mean longitudes." We now define it as a celestial orb's angle of tilt with the ecliptic.

Latitude, celestial "Measured from 0 deg. to 90 deg. at right angles to the ecliptic, along a circle passing through the object and the ecliptic poles." *Not* a projection of terrestrial latitude on the celestial sphere, which is **declination**.

Libration [Copernican]. "Two reciprocal movements belonging wholly to the poles" of a celestial orb (esp. the Earth), "like hanging balances."

Lightyear The distance light travels in a year, 9.46×10^{12} kilometers.

Longitude, celestial "Measured along the ecliptic from 0 deg. to 360 deg. east of the **vernal equinox**." *Not* a projection of terrestrial longitude on the celestial sphere, which is **declination**.

Loxosis [Ptolemaic]. Rocking of a planet within its theoretical path of orbit.

Magnitude Brightness of a celestial body, now measured logarithmically.

Meridian The plane containing the observer on a given planet and that planet's rotational axis.

Meridian, celestial The projection of a terrestrial observer's own meridian upon the celestial sphere. It passes through his zenith and his horizon's north and south.

Node One of the points where a moon or planet's orbital plane crosses the ecliptic.

Obliquation [Copernican]. A "small periodic fluctuation in the inclinations of the planetary deferents." Copernicus introduces obliquation (VI.1) as the latitude at the highest and lowest **apsides** of a celestial orb.

Obliquity The angle between a celestial orb's orbital plane and its equator.

Obliquity of the ecliptic The angle between the Earth's equator and the ecliptic plane. About 23° 27'. If a celestial orb other than Earth is referred to, then its obliquity *vis-à-vis* the ecliptic is obviously the angle between *its* equator and the ecliptic plane.

Occultation Disappearance of a smaller body behind a larger. Jensen reminds us: "The body need only be smaller in angular size. Often the moon occults a star; of course the star is in reality much larger than the moon, but it is also much farther away, so its angular size (or apparent size) is smaller.

Opposition When a celestial body and the sun appear to be separated from each other by 180 degrees in the terrestrial sky. *See* **inferior conjunction.**

Orbit The rotational path of a celestial orb about a central body. In case the theme of this book hasn't yet become apparent to you, Ptolemy thought that our neighboring planets orbited about the Earth, while Copernicus opined that they orbited the sun. Ptolemaic and Copernican orbits were comprised of as many circles as it took to create a perfect circle in conformity with "the appearances." We now know that the planetary orbits are characterized by **eccentricity.** In fact, they are elliptical.

Parallax The changing angle in relation to a background object of a given foreground object as our point of reference moves. It was argued against the Copernican theory that there should have been stellar parallax if the Earth really did move around the sun. There was, but the stars were so much farther away than everyone thought that stellar parallax remained immeasurable for centuries after *Revolutions.* Planetary parallax was in fact measurable, and helped Copernicus to explain the appearances.

Period, orbital The amount of time required to complete one orbit.

Period, sidereal The time it takes for one celestial body to complete a revolution around another (Jensen adds here: "as measured with respect to the stars"). For the Earth, a sidereal year is 365 days, 6 hours, 9 minutes, 9.3 seconds.

Period, synodic The time it takes for one celestial body to return to the same apparent phase around another with which it began. For the moon, one sidereal period is 27 days, 7 hours, 43 minutes, 11.5 seconds; its synodic period is 29 days, 12 hours, 44 minutes, 2.8 seconds, since the sun's apparent position has altered eastward in the meantime.

Perigee The point in the orbital path where a celestial orb is nearest to the Earth.

Perihelion The point in the orbital path where a celestial orb is nearest to the sun.

Pole, celestial See **sphere, celestial.**

Precession The slow wobbling of the Earth on its axis, due to solar and lunar gravitational pull. A complete precessional period is about 25,800 years.

Retrograde Condition when an aparent planetary motion around the earth seems to reverse.

Quadrature When a celestial body and the sun appear to be separated from each other by 90 degrees in the terrestrial sky.

Right ascension A projection of terrestrial longitude on the celestial sphere.

Secular Taking place over a very long time.

Solstice, summer The point on the ecliptic closest to the northern celestial pole; also, the day of the year when the sun rises and sets most to the north, and the day is the longest.

Solstice, winter The day of the year when the sun rises and sets in the most southerly direction, and the night is longest. Also, the point on the ecliptic farthest from the northern celestial pole.

Sphere, celestial An imaginary sphere, centered around our point of reference on the Earth, which contains all the celestial objects which we can see. The Earth's poles and equator are projected onto their celestial counterparts.

Star, fixed [Copernican and pre-Copernican]. Any star except a "wandering star" (= a planet), because all stars were thought to be mounted (fixed) on the outermost sphere of the universe, which revolved around us every twenty-four hours.

Sublunar (or sublunary) realm [Copernican and pre-Copernican]. The terrestrial zone beneath the Sphere of the Moon. Aristotelians claimed that this realm, being composed of the four elements, is subject to mortality, corruption and change; while the realm above the moon, the superlunar or superlunary spheres, was filled with an imperishable fifth element, ether. The heavenly bodies were similarly imperishable.

Synodic The interval "between successive identical phases of a planet or satellite."

Transit A crossing by one celestial body of another, as seen by an observer in a third place. In this book the most relevant case is the transit of Venus across the face of the sun in 1639.

Trepidation [Ancient astronomy, in place of idea of precession; and Copernican, combined with precession]. "Periodic fluctuation of the equinoxes and solstices, resulting from an oscillation of the inclination of the ecliptic."

Tropics [Copernican and pre-Copernican]. The two latitude lines on the celestial sphere (i.e., parallels to the celestial equator) which the apparent sun crosses at its farthest north and farthest south of the ecliptic (± 23° 27'). Also, the two corresponding terrestrial latitudes (23° 27' N and 23° 27' S).

World [Copernican and pre-Copernican]. The cosmos.

Year, sidereal See **period, sidereal.**

Zenith The point on the **celestial sphere** directly overhead from the observer.

Sources

I have repunctuated and "retranslated" freely whenever I thought I could improve sense or style. For example, Galileo's ". . . in the case of the other planets I shall try—with the help of God, without Whom we can do nothing—to make a more detailed inquiry concerning them . . ." has been robbed of both ellipses: "In the case of the other planets I shall try—with the help of God, without Whom we can do nothing—to make a more detailed inquiry concerning them." Copernicus's "What could be more beautiful than the heavens which contain all beautiful things?" actually begins "For what . . . ," and once again I omitted the ellipsis. *Tant pis.* Sometimes, however, ellipses are indicated in the citations below. Keywords in quotation marks are from cited material. Unadorned keywords come from my own text or from a paraphrase.

p. 7:   Epigraph—Hartmann, p. 2.

p. 19:   "By the name of Heaven and Earth"—Augustine, p. 104 (*Confessions*, XII.17). My translation capitalizes neither "heaven" nor "earth," so once I capitalized "Earth" it would have distorted Augustine's faith not to capitalize "Heaven," also; see fn. to *Confessions*, XII.7, below.

p. 21:   Vatican astronomer: "His tables were superseded"—Consolmagno, p. 62.

p. 21:   "HE WAS A SCHOLAR OF POLISH BIRTH"—Frontispiece, Copernicus Anniversary National Organizing Committee.

p. 22:   "How great the orbital circles of Saturn"—Copernicus, p. 818 (VI.3).

p. 22:   So why browse through *Revolutions*'s mathematics-cobwebbed pages?—I propose to read its arguments with you in some detail, but not too much, because so many of them have also worn out that it would be pointless, not to mention sadly tedious, to crumble away every last moth-eaten geometric proof.

p. 22:   Newton's First Law—Tilley and Thumm, p. 57.

p. 22:   Aristotle as one of the first proponents of a spherical Earth—Kopal (p. 24) even claims him to be the very first.

p. 23:   Jungians: "Collective symbols of the Self"—Von Franz, p. 24.

p. 23:   Forty-eight epicycles for the paths of the five known planets—Koestler, p. 124.

p. 24:   "Our problem is to find arc *FC*"—Copernicus, p. 810 (V.36).

p. 25:   *"What could be more beautiful than the heavens"*—Copernicus, p. 510 (I, Osiander's preface); initial "For" omitted.

p. 25:   "The first movement of approach towards that state of mental purity"—Herschel, p. 2.

p. 26:   "For the motion of a sphere is to turn in a circle"—Copernicus, p. 513 (I.4). Italics mine.

p. 26:   "It would be useless to deny the existence"—Dugas, p. 18.

p. 26:   Descriptions of the two Martian moons' shape and surface variations—Based on Greeley and Batson, pp. 125–26 ("Mars System: Satellites").

p. 28:   "For according to Saint Dionysus"—Ross and McLaughlin, p. 235 (Pope Boniface VIII, from the Bull "Unam Sanctam," excerpted here as "The Superiority of the Spiritual Authority").

p. 28:   "The poor laborer must needs pay the wages"—Coulton, p. 325 (États Generaux de Tours, *Recueil,* pp. 88ff).

p. 28:   Desperate gleaners are selling themselves—Ibid., p. 328 (citing Werner Rolewink).

p. 28:   Similar conditions on Poland—Milosz, p. 26: "At first the peasants were required to work one day a week for their lord, then two, then three, and so on, until their lot was reduced to a state of enslavement."

p. 29: Preface and Copernicus's fatal stroke—Copernicus Anniversary National Organizing Committee, p. 11.

p. 29: Addition *of Heavenly Spheres* to the title by Osiander—Ibid., p. 10.

p. 29: "Osiander's preface has maintained the ambiguity"—Dugas, p. 37.

p. 29: "The world" is a sphere "because this figure is the most perfect of all," etc.—Copernicus, p. 511 (I.1).

p. 30: "And no one would hesitate to say that this form belongs to the heavenly bodies"—Loc. cit.

p. 30: Our moon is "highly asymmetric"—Taylor, p. 106.

p. 30: Copernicus's twenty-seven recorded observations—Jacobsen, p. 140.

p. 31: "If I could bring my computations to agree with the truth"—Ibid., p. 105.

p. 31: "Since the mind shudders"—Copernicus, p. 514 (I.4).

p. 32: "Today this intense faith in number harmonies"—Kuhn, p. 219.

p. 32: "It is easy to speak of the infinite . . . but . . . difficult to speak of it meaningfully."—Quoted in Hartmann, p. 413.

p. 32: "Such a sphere, of unlimited radius"—Ayres, p. 199.

p. 33: "Among our principles and hypotheses"—Copernicus, p. 585 (II.14).

p. 35: Spheres eventually came to serve more as constructs—Kuhn, p. 105.

p. 35: "Rather deep so as not to be the atmosphere"—Ptolemy, p. 291 (VIII.3).

p. 35: "In yellow or some other distinct color."—Ibid., p. 292 (loc. cit.).

p. 35: "Many however have believed that they could show"—Copernicus, p. 515 (I.5).

p. 36: "Dilating" one's thoughts "to comprehend the grandeur"—Herschel, pp. 2–3.

p. 36: "And therefore out of nothing"—Augustine, p. 101 (*Confessions*, XII.7). My translation capitalizes neither "heaven" nor "earth," so once I capitalized "Earth," which is "a small thing," I had to also capitalize "the great thing," namely "Heaven," in order to avoid an orthographic absurdity.

p. 36: Remarks on Guido Bonatti—From Seligmann, p. 135.

p. 36: Jewish belief in an allotted star—Trachtenberg, p. 250.

p. 37: "I know of no one who doubts"—Copernicus, p. 521 (I.10) ("heavens of the fixed stars" in original).

p. 37: "Let AB be the greatest circle"—Ibid., p. 653 (III.15).

p. 37: "It seems more accurate to say that the equator"—Ibid., p. 623 (III.1).

p. 37: Qur-'An cited for a spherical Earth—Al-Biruni, pp. 23–24. The surah cited is 11;7.

p. 38: By fourth century B.C., Greek thinkers most often asserted a spherical Earth—Kuhn, p. 27.

p. 38: Herschel on Earth's orange shape—Op. cit., p. 130. According to Kopal (p. 24), it is "a rotational spheroid" slightly flatter at both poles than elsewhere.

p. 38: "In finding the distance between two points"—Welchons and Krickenberger, p. 329.

p. 39: Description of dip sector—Herschel, p. 17.

p. 39: Infrared photo of Earth's curvature—Kuiper 1954, p. 714 (Clyde T. Holliday, "The Earth as Seen from outside the Atmosphere").

p. 39: Ptolemy's second observational proof—Op. cit., p. 9 (I.4).

p. 39: "The northern vertex of the axis"—Copernicus, p. 511 (I.2). I have omitted the following more obscure proof: The inclinations of the poles "have everywhere the same ratio with places at equal distances from the poles of the earth, and that happens in no other figure except the spherical."

p. 40: Noon shadow points north—Kuhn, pp. 8–9.

p. 40: "Add to this the fact"—Copernicus, p. 512 (I.2).

p. 40: Earth is spherical since its shadow on the moon is circular—This argument is actually made at the end of the next section. Copernicus, p. 513 (I.3).

p. 40: "So Italy does not see Canopus"—Ibid., p. 512 (I.2).

p. 41: "Spheres are to one another as the cubes"—Loc. cit. (I.3).

p. 41: Water covers almost 75 percent of Earth's surface—Greeley and Batson, p. 81.

p. 41: Oceans are five miles deep—Wells, p. 16.

p. 41: 1/792 of the planetary radius—Taylor, p. 92, gives the figure of 3726.8 miles at the equator (6378 kilometers); but for the sake of consistency I will once more employ the figure of 3960 miles given by Welchons and Krickenberger.

p. 42: "We would not be greatly surprised if there were antipodes"—Copernicus, p. 513 (I.2).

p. 42: "We will recall that the movement of the celestial bodies is circular."—Ibid., p. 513 (I.4).

p. 42: "Where the axis of the Earth is perpendicular"—Ibid., p. 566 (II.5).

p. 43: "The directions are the same."—Jensen, n. to ms. p. 41.

p. 43: "That by which everything moves from east to west"—Ptolemy, p. 12 (I.8).

p. 44: Sentence beginning "Copernicus has already informed us"—"The above description is only strictly true of stars on the celestial equator," Eric Jensen reminds me (n. to ms. p. 42). All the same, even the North Star moves fifteen degrees per hour around the North Celestial Pole, even if its motion is almost indiscernible to us.

p. 44: "Make certain complex movements"—Ptolemy, p. 13 (I.8).

p. 44: "They exhibit initiative by their individual course"—Seligmann, p. 249.

p. 44: "To suppose a second movement"—Ptolemy, loc. cit.

p. 44: "The line on which eclipses may occur."—Gleadow, p. 17.

p. 45: Eight or nine degrees to each side—Aaboe (p. 6) allows slightly more space to the whole contrivance, remarking that the moon and planets are never more than ten degrees off the ecliptic.

p. 45: Two theories to explain approximate coincidence of orbital planes—Whipple, p. 3.

p. 45: Second theory has won out—Jensen, n. to ms. p. 44.

p. 45: Twelve zones of exactly thirty degrees each—Gleadow, p. 16.

p. 45: *Dodekatemoria*—E.g., Copernicus, p. 576 (II.9).

p. 45: Their king risked being stung to death by a scorpion—Seligmann, p. 6.

p. 46: Rotation of the celestial equator a result of solar and lunar gravitation—Hartmann, pp. 12, 36.

p. 46: Copernicus's calculation of the precessional period 99.9 percent accurate—Copernicus Anniversary National Organizing Committee, p. 15: Copernicus's calculated revolution of the polar axis was 25,809 years. The modern value is 25,785 years.

p. 47: "The equinoxes of which the one guarding the northern approach"—Ptolemy, loc. cit.

p. 47: At equinoxes, day and night become equal in length—Somewhat after Aaboe, pp. 5–6.

p. 47: Two points in time become magically transmuted into a pair of points in space—And more than two. Ptolemy then invites us to draw a new great circle which cuts both the equator and the ecliptic in half. The two points where the new great circle crosses the ecliptic will be the winter and summer tropics.

p. 47: Days and nights become proportions of a circle—As implied by Copernicus, p. 576 (II.8).

p. 47: "The duration of days and nights are inversely equal"—Ibid., p. 569 (II.6).

p. 47: Copernicus insists that the variable obliquity of the ecliptic should really be thought of as the variable inclination of the equator—Ibid., p. 567 (II.6).

p. 48: Ecliptic wriggles on 10 June 2004 and other dates—In this paragraph and the one above which describes the sun's and planets' motions as seen from Earth, I am making use of the "Starry Night Pro" astronomical software package (2000; version 3.1), which allows one to simulate a view of the sky from any point on Earth, at any time; and also to gaze down on the solar system and observe the paths of the planets from any angle.

p. 49: "As the ecliptic is oblique to the axis"—Copernicus, p. 577 (II.10).

p. 50: Sun's apparent motion is fastest in January and slowest in July—Jacobsen, p. 7.

p. 50: "Each day the sun moves rapidly westward *with the stars*"—Kuhn, p. 24.

p. 50: "The great circle in which the plane of the Earth's orbit"—Weigert and Zimmermann, p. 99.

p. 51: Earth's orbital plane passes through the line between the center of gravity, etc.—Loc. cit.

p. 51: Precession at fifty arc-seconds per year—Gleadow, p. 19.

p. 51: "The Eighth Sphere's motion, which the ancient astronomers"—Copernicus (Rosen), vol. 3, p. 147 (letter against Werner).

p. 52: "For the sun and moon are caught moving"—Copernicus, p. 514 (I.4).

p. 52: "And we perceive the five wandering stars sometimes even to retrograde"—Loc. cit.

p. 52: "They maintain these irregularities"—Loc. cit.

p. 52: "It is agreed that their regular movements"—Loc. cit.

p. 52: "Attribute to the celestial bodies what belongs to the Earth."—Loc. cit.

p. 52: Copernicus, first to explain the retrograde motions of the planets—Jacobsen, p. 110.

p. 53: "Variety usually gives greater pleasure"—Copernicus (Rosen), vol. 3, p. 29 ("A Letter of Nicholas Copernicus to the Right Reverend Lucas [Watzenrode], Bishop of Varmia").

p. 53: "The epoch in which Copernicus lived"—Milosz, p. 37.

p. 54: "For they kept seeing the sun and moon"—Ptolemy, p. 7 (I.3).

p. 55: On Venus the sun rises in the west—Kopal, p. 51.

p. 56: Copernicus crosses out Aristarchus's name—The fact and its possible interpretation are both from Jacobsen, p. 108. Rosen disputes this motive in *Copernicus and His Successors* (pp. 4, 8), admitting the deletion but explaining it as the result of ignorance: "Poor Copernicus, his imperfect sources led him to confuse Aristarchus with Eratosthenes and Callipus." I don't quite see this. It would be one thing if Copernicus never mentioned Aristarchus at all. But to mention him and then cross him out is hard to explain away on the basis of ignorance. Meanwhile, the issue gets further obfuscated by the fact that Copernicus does mention Aristarchus elsewhere in *Revolutions*.

p. 56: Aristarchus was the first—According to Rosen (ibid., p. 10).

p. 56: "From him, Copernicus adopted the idea"—Copernicus Anniversary National Organizing Committee, p. 4.

p. 57: Heliocentrism "mentioned by at least twelve philosophers"—Jacobsen, p. 107.

p. 59: "How the Earth moves"—Ross and McLaughlin, p. 623 (Abelard of Bath, *Questiones naturales*, early twelfth century).

p. 59: "Its perfection is such"—Ptolemy, p. 1 (introduction by R. Catesby Taliaferro).

p. 59: What then were Ptolemy's arguments?—Let's omit an argument about the timing of lunar eclipses. It's not only wrong; it's gruesomely technical.

p. 59: At solstice, water was said to become poisonous—Trachtenberg, p. 257.

p. 60: The creator "made seven unequal circles"—Plato, p. 1166 (*Timaeus*, 36d).

p. 61: Because the stars twinkle and the planets do not—Copernicus (Rosen), vol. 3, p. 143 (letter against Werner).

p. 61: "At this time Canicula was beginning to rise"—Copernicus, p. 643 (III.10).

p. 61: "The longest tyranny"—Copernicus Anniversary National Organizing Committee, p. 6.

p. 62: New moon, "first quarter"—Jacobsen, p. 8.

p. 63: "Independent discovery is quite out of the question,"—Aaboe, p. 109.

p. 63: "I must emphasize that once"—Ibid., p. 111.

p. 63: Relative celestial diameters table: Ptolemy's calc.—Ptolemy, p. 176 (V.16).

p. 63: Same table; Copernicus's calc.—Copernicus, p. 713 (IV.20). The sun's diameter is given as eighteen times that of the Earth.

p. 63: Same table: Current moon est.—Greeley and Batson, p. 325: Lunar radius = 1737.4 kilometers, so diameter = 3474.8 kilometers.

p. 63: Same table: Current Earth est.—Loc. cit.: Earth's diameter is 6378.14 kilometers.

p. 63: Same table: Current sun est.—Ibid., p. 30: The sun is "more than 1,390,000 kilometers across."

p. 64: Duration of winter and summer in northern hemisphere—Copernicus Anniversary National Organizing Committee, p. 5. Jacobsen prefers the respective Figs. 179 and 186.

p. 64: "The Earth in its annual revolution"—Copernicus, p. 653 (III.15).

p. 65: Ptolemy's Prime Movement—Ptolemy, p. 233 (VII.4).

p. 66: Ptolemy and his predecessors—Aaboe, p. 72, remarks that the epicyclic model is "of unknown origin." Some authorities credit Eudoxus.

p. 66:   "A simple epicyclic model"—Aaboe, p. 76.

p. 66:   "It is first necessary to assume in general that the motions"— Ptolemy, p. 86 (III.3).

p. 66:   "Considered with respect to a circle"—Ibid., p. 87 (loc. cit.).

p. 67:   Eastward one degree per century—Ibid., p. 292 (IX.6).

p. 68:   "The eccentric circle [or deferent, which"—Ibid., p. 322 (X.6).

p. 69:   Anomalies of the planets with respect to sun and Zodiac—Ibid., p. 291 (IX.5). Here Ptolemy writes "stars" but means "wandering stars."

p. 69:   "The ancients found in all the planets"—Copernicus, p. 813 (VI.1).

p. 69:   Both Copernicus and Ptolemy see lunar orbit as epicycle—Ibid., p. 526 (I.10).

p. 69:   *Revolutions*'s account of the lunar orbit—Jacobsen, p. 133. A helpful diagram appears here.

p. 69:   Plato's eight diversely colored celestial spheres—Plato, pp. 840–41 (*Republic*, X, 616b–17d).

p. 69:   "In Ptolemy's universe, mathematics"—Taub, p. 153.

p. 70:   "This brilliant device"—Sir Bernard Lovell, p. 29.

p. 70:   Indeed, with these circles we can do more than that—The discussion in this paragraph is heavily indebted to Ptolemy's translator's explanation (Ptolemy, pp. 470–72, "Appendix B: The Passage from the Ptolemaic to the Copernican System, and Thence to That of Kepler").

p. 70:   Radius of Ptolemy's Martian epicycle and eccentric circle—Op. cit., p. 342 (X.8).

p. 70:   Mean Martian orbital radius of 1.524 AU—Kopal, p. 55.

p. 70:   Hipparchus's delicate little epicycle—Sir Bernard Lovell, pp. 27, 30. This epicycle revolves two times westward for each eastward revolution of the deferent; Lowell informs us that such a configuration will adequately model the sun's irregular apparent motion around us.

p. 71:   The mechanical drawing, *circa* 1615—Lefèvre, pp. 24 (Fig. 1.5, Caus 1615, fol. 24r), 23 (Fig 1.4, ibid., fol. 23r). Fig 1.5 is the view which I am considering most; Fig. 1.4 is the perspective view. The question of whether Plato considered ether to be a physical element as Aristotle did (Xenocrates thought he did, Cicero that he didn't) applies to Ptolemy as well.

p. 71: "The epicycles' centers are borne on circles"—Ptolemy, p. 291 (IX.5). Italics mine.

p. 72: Now the motion of the planet is unvarying with respect to some new point—I am indebted to Kuhn, p. 71, for this definition.

p. 74: *Oscillates around a mean position*—The italicized phrase is Aaboe's.

p. 74: "This broadening of the principle of celestial mechanics"—Ptolemy, p. 292 (fn. to IX.5).

p. 74: "And so there were three centers"—Copernicus, p. 785 (V.25).

p. 75: "Possible to imagine a variable speed"—Al-Biruni, p. 28.

p. 75: "His results were for the most part no more accurate"—Jacobsen, p. 111.

p. 76: "And so we shall arrange the table"—Ptolemy, p. 102 (III.6).

p. 76: "Whenever we wish to know the course of the sun"—Ibid., p. 104 (III.8).

p. 76: The center of the sun's epicycle speeds up and slows down—Ibid., p. 105 (III.9).

p. 77: "The angle at B"—Ibid., p. 323 (X.6).

p. 77: "The courses of the planets"—Copernicus, p. 511 (Book I, introductory remarks).

p. 78: Footnote on Albertus Magnus—Weisheipl, pp. 164–65, 175.

p. 78: "We should rather follow the wisdom of nature"—Copernicus, p. 526 (I.10).

p. 79: "Now that it has been shown that the Earth"—Ibid., p. 514 (I.5).

p. 79: The Newtonian mechanics we now *routinely* utilize—This qualification is a concession to Einstein's relativity theory.

p. 80: Each of the four seasons partook of its own element—Bede, pp. 100–101.

p. 81: "Blood, which increases"—Loc. cit.

p. 81: Saturn was banefully cold and dry—Weisheipl, pp. 176–77 (Betsey Barker Price, University of Toronto, "The Physical Astronomy and Astrology of Albertus Magnus").

p. 81: "Take musk, ambergris"—Barret, p. 93.

p. 82:   At 10,000 feet, the same volume of air—Kotsch, p. 289 (Appendix A, Table A-6, "Altitude and Atmosphere Table").

p. 82:   "Has *within itself*"—Aristotle, p. 268 (*Physics*, II.1.192b). Of course I am grossly simplifying Aristotle's views, which altered over his career. For instance, in *Of Generation and Corruption* he proposes that the four elements can metamorphose into each other, whereas in *On the Heavens* they cannot.

p. 82:   "Water, which by its nature flows"—Copernicus, p. 512 (I.2).

p. 82:   "Earth is the heaviest element"—Ibid., p. 517 (I.7).

p. 82:   "Since Copernicus does not understand physics"—Quoted in Rosen, pp. 158–59.

p. 83:   "What exists in a state of fulfillment only"—Aristotle, p. 278 (*Physics*, III.I.200a).

p. 83:   "The energy that a system possesses"—Tilley and Thumm, p. 158.

p. 84:   A few simple algebraic transformations allow us to determine the mass of the Earth—Ibid., p. 67. The equation is: $T^2 = (4\pi^2/GM)r^3$, where T = the period, G = the gravitational constant, M = the mass of the central body about which the other one orbits, r = the orbital radius. Jensen advises (n. to ms. p. 86) that "in place of M we should of course write $(m + M)$, the small-case m being the orbiting body's mass. When the central body is the sun, the sum is like adding the weights of a flea and an elephant . . . Even Jupiter is only 1/1000 of the sun's mass. So having just M instead of $M = m$ isn't that far off," although it could be if (for instance) a larger than usual moon orbited a smaller than usual planet.

p. 84:   "There is no such thing as motion"—Aristotle, p. 278 (*Physics*, III.I.201a).

p. 84:   "It is always with respect to substance or to quantity"—Ibid., p. 294 (*Physics*, IV.8.215a).

p. 84:   Each element has its own natural law, its own "virtue"—Elders makes this point in regard to Aristotle when he says that "the point of departure" for this philosopher "is an intuition of an essence or an essential law of a natural body."

p. 84:   The motion of water being raised in a bucket is compulsory—Aristotle, p. 295 (216a). I am leaving out many aspects of Aristotle's theory of motion which correspond to our own. For instance, Aristotle asserts, as would we, that motion is affected by the degree of drag of the medium through which the moving object travels.

p. 84: "A simple body cannot have two natural motions"—Quoted in Rosen, p. 156.

p. 85: "It is the fulfillment of what is potential"—Aristotle, p. 279 (III.I.201b).

p. 85: Empedocles supposed Love to be the moving force of the cosmos—Elders, p. 20.

p. 85: "Here power failed the high phantasy"—Dante, *Paradiso*, p. 485 (end of Canto XXXIII).

p. 86: "It is impossible that a simple heavenly body should be moved"—Copernicus, p. 514 (I.4).

p. 86: He defines uniform motion as *mean motion.*—Copernicus (Rosen), vol. 3, p. 148.

p. 86: "In their investigations of the moon's path"—Ibid., p. 147.

p. 87: "Circular movement always goes on regularly"—Copernicus, p. 520 (I.8).

p. 87: "The principles of this art might be preserved"—Ibid., p. 740 (V.2).

p. 88: "Note that we do much the same"—Jensen, n. to ms. p. 90.

p. 89: "*The innermost motionless boundary*"—Aristotle, p. 291 (*Physics*, IV.4.212a). Italics in original.

p. 89: "*Heaviness is a body's resistance*"—Plato, *Timaeus*, 60c.

p. 90: "Every apparent change in place"—Copernicus, p. 514 (I.5).

p. 91: "For the daily revolution appears to carry the whole universe"—Ibid., p. 515 (I.5). Italics mine.

p. 91: "It is not necessary that these hypotheses"—Ibid., p. 505 (Osiander's preface).

p. 91: "The fact that the wandering stars"—Ibid., p. 515 (I.5).

p. 92: "And God made the firmament"—Genesis 1.7–8.

p. 92: "Now let the horizon be the circle *ABCD*"—Copernicus, p. 516 (I.6).

p. 93: "Therefore, since *AEC* is in a straight line"—Ibid. Likewise all following quotations in this paragraph.

p. 93: "From this argument"—Ibid.

p. 93: Objects within the solar system are another matter—Norton, p. 3 ("Position": "Celestial Coordinate Systems").

p. 93: Mariners need not apply parallax corrections—Davies, p. 109. The moon is an exception.

p. 94: "But we see that nothing more than that"—Copernicus, loc. cit.

p. 94: "It is necessary that movement around the center"—Ibid., p. 520 (I.8).

p. 94: "All the more then will the Earth"—Ibid., p. 517 (I.7).

p. 94: "I myself think that gravity"—Ibid., p. 521 (I.9).

p. 95: Why don't the inhabitants of the Sphere of Fixed Stars fly apart into pieces?—Ibid., p. 518 (I.8).

p. 95: "Small steel balls of radius 10^{-4} meter"—Tilley and Thumm, pp. 216–17.

p. 96: The Neptunian winds blow east to west—Greely and Batson, p. 293.

p. 97: The Coriolis Effect is moderated—This exposition is based on the discussion in Kotsch, pp. 39–42, 87–89.

p. 97: On an ocean cruise, a ball thrown straight up—Herschel, p. 14.

p. 98: "Then what should we say about the clouds"—Copernicus, p. 519 (I.8).

p. 99: "Lastly, the sun will be regarded"—Ibid., p. 521 (I.9).

p. 100: But most eyes cannot distinguish stars—Norton, p. 23.

p. 100: Either system went far to "save the appearances"—Copernicus Anniversary National Organizing Committee, p. 11.

p. 100: "Let us permit these new hypotheses"—Copernicus, p. 506 (Osiander's preface).

p. 101: "The rocket photographs show"—Kuiper 1954, p. 715 (Clyde T. Holliday, "The Earth as Seen from outside the Atmosphere").

p. 101: By means of a wire plus the following para.—Description after Herschel, pp. 156–58. Jensen didn't care for Herschel's description (or my rewording thereof) of the cause of Foucaultian motionlessness at the equator—Earth's rotation vectors point directly perpendicular to the ground at the equator—so I cut that out.

p. 102: $15 \times 0.7528 = 11.292$—I took the sine of Paris's latitude from Welchons and Krickenberger, end matter p. 73 (Table X, "Natural Functions and Radians," which breaks down into ten-minute intervals—good enough for government work).

p. 102: 1674 kilometers per hour—Deligeorges, p. 4.

p. 103: Copernicus never observed Mercury—Copernicus Anniversary National Organizing Committee, p. 15.

p. 103: He laments the obscuring vapors of the Vistula—Copernicus, pp. 793–94 (V.30): "For nature has denied that convenience" of clear skies "to those of us who inhabit a cold region where fair weather is rarer . . . On this account the planet has made us take many detours and undergo much labor in order to examine its wanderings. On this account we have borrowed three positions from those which have been carefully observed at Nuremberg."

p. 103: "The general experience" of searching for Mercury—Norton, p. 72.

p. 103: "Partly a result of illumination conditions"—Taylor, p. 155. All descriptions in my sentence above are based on plates in this book.

p. 104: Wood might warp and mislead us—Copernicus, p. 558 (II.2).

p. 104: "Saturn, the highest"—Ibid., p. 517 (I.6).

p. 104: "Ptolemy marked the greatest southern latitude"—Ibid., p. 817 (VI.3).

p. 104: Copernicus puts it at 240° 21'—Rosen, p. 36.

p. 104: His instruments are the same as Ptolemy's—Bienkowska, p. 23.

p. 104: "He made some observations"—Weigert and Zimmermann, p. 59 (entry on Copernicus).

p. 104: Ptolemy's value for the maximum angular elongation of Venus— Ptolemy obtains a very respectable greatest eastern elongation from the true sun of 46° 22' in the sign of the Ram (op. cit., p. 422 [XII.9]). He also provides a table of various western and eastern elongation values for this "star." The minimum value is an eastern elongation of 44° 25' in the sign of the Crab; the maximum is a western elongation of 46° 47' in the Scorpion (ibid., p. 424 [loc. cit.]). Our modern value is about 47°.

p. 105: "The work for another's love of wisdom and truth"—Ptolemy, p. 81 (III.2).

p. 105: "Hold fast to their observations"—Copernicus (Rosen), vol. 3, p. 147 (letter against Werner).

p. 105: "Since it was not permissible to ignore them"—Koestler, p. 128.

p. 105: "But neither 3 minutes nor 4 minutes"—Copernicus, p. 830 VI.7). Italics mine.

p. 105: "The dark spots on the moon"—Lear, p. 171 (*The Dream,* by Johannes Kepler, the Late Imperial Mathematician: A Posthumous Work on Lunar Astronomy [Sagan in Silesia, 1634; written in or after 1608]).

p. 106: Description of that unassuming brick structure—Bietkowski and Zonn, unnumbered p., plate 150.

p. 106: "Wander in various ways"—Copernicus, p. 514 (I.4).

p. 106: "Watching their movement from night to night"—Ridpath, p. 15.

p. 106: "Binoculars are usually needed to spot it"—Ibid., pp. 32–33, 47, 43, 51, 30.

p. 107: His faith in erroneous observations introduced contradictions, leading him to reject Ptolemy's system—Kuhn, pp. 140–41.

p. 107: "We are approaching the limits"—Hartmann, p. 413.

p. 107: "The subsequent 23 years"—Jensen, n. to ms. p. 117.

p. 107: "And as far as hypotheses go"—Copernicus, p. 506 (Osiander's preface).

p. 108: "Thus the heliocentric system"—Copernicus Anniversary National Organizing Committee, p. 22.

p. 109: "The magnitude of the orbital circles"—Copernicus, p. 526 (I.10).

p. 110: "It is necessary that the space"—Ibid., p. 525 (I.10).

p. 110: "Therefore we are not ashamed to maintain"—Ibid., p. 525 (I.10).

p. 110: "Copernicus has telescoped the eccentric circle"—Ibid., p. 528 (continuation of fn. 1 from p. 526).

p. 110: "I also say that the sun remains"—Ibid., p. 525 (I.10).

p. 111: "So it happens that the sun itself"—Ibid., p. 530 (I.11).

p. 111: "It follows that the axis of the Earth"—Ibid., p. 530 (I.11).

p. 111: Equinoxes and solstices have altered by twenty degrees—Ibid., p. 523 (I.11)

p. 111: "The ecliptic remains perpetually unchangeable"—Ibid., p. 626 (III.3).

p. 111: Before him, astronomers had thought that it went the other way.—Jacobsen, p. 127.

p. 112: "Since we see that we have come so far"—Copernicus, p. 538 (I.12).

p. 113: Venus, which to the Mesopotamians was fertile Ishtar—Seligmann, p. 6.

p. 113: Helps men to win the hearts of women—Barret, p. 163.

p. 113: "A veritable inferno"—Kopal, p. 48.

p. 113: Venus's retrograde rotation shared only by Uranus—Ibid., p. 50.

p. 113: 584 days between inferior conjunctions—Norton, p. 73.

p. 114: Certain coincidences of angular velocity—Kopal, p. 51. Before Ptolemy, Eudoxus with his homocentric spheres (the simpler precursors of epicycles and deferents) twiddled his inner and outer spheres in various fashions until he came up with an explicatory bowtie-like figure, called the hippopede, a horse-hobble. For Venus we have 5 synodic cycles in 8 years or about 8 revolutions in the ecliptic of Venus. "In terms of Eudoxos's homocentric sphere," writes Aaboe in his history, "Venus must then travel through its hippopede 5 times in 8 years, and the hippopede must be carried 8 times in the ecliptic in the same 8 years" (Aaboe, p. 71). It's all so clever, but Eudoxus's hippopedes cannot represent those mischievously retrograde moments, when Venus seems to orbit backwards. At least Ptolemy finds a way to let Venus go retrograde near the epicycle's perigee (ibid., p. 79).

p. 114: Astrologically favorable or not, depending on sympathy or contradiction between the orbs—Seligmann, p. 252.

p. 114: "Now all planets are afraid"—Barret, p. 149.

p. 114: "In the fifth place Venus"—Copernicus, p. 527 (I.10).

p. 114: One revolution of Venus, 224.7 days—Norton, p. 73.

p. 114: Radius of the Venusian orbital circle is 7193—Copernicus, pp. 778–79 (V.21).

p. 114: Mean radius of the Venusian orbit = 0.7233 × Earth's—Kopal, p. 23.

p. 114: "All motions in the diagram"—Jacobsen, p. 133.

p. 115: Copernicus "seems to feel that he has won his case"—Vermij, p. 34

p. 115: "And in the year 21 of Hadrian"—Ptolemy, p. 312 (X.1).

p. 115: Then he draws three overlapping circles—Ibid., p. 316 (X.3).

p. 115: "On the 4th day before the Ides of March"—Copernicus, p. 782 (V.23).

p. 116: The sun would sometimes "be eclipsed in proportion"—Ibid., p. 522 (I.10).

p. 116: Mercury, the Mesopotamian god of wisdom—Seligmann, p. 5.

p. 116: "Too small to see without a telescope."—Hartmann, p. 106.

p. 117: No "sensible parallax"—Ptolemy, p. 270 (IX.1).

p. 117: There is *always* parallax—Copernicus, p. 734 (fn. to V.1).

p. 117: Footnote: The "astronomical triangle" for the heavenly body T—Ayres, pp. 199–200.

p. 118: "I'll say!"—Jensen, n. to ms. p. 131.

p. 118: "It is clear that in the case of those stars"—Ptolemy, p. 165 (V.11).

p. 118: Copernicus informs us how to construct a parallacticon—Copernicus, pp. 705–706 (IV.15).

p. 119: Ptolemy's derived Earth-moon distance—Ptolemy, pp. 167–71 (V.13).

p. 119: Footnote: "This seems to be mixing apples and oranges"—Jensen, n. to ms. p. 132. In Ptolemy, p. 139 (IV.10), it is explicitly stated that the ratio of 3144 to $327\frac{2}{3}$ = the ratio of 60 theta to 6 theta 15'. If the latter figure is considered to be 6.25 theta, then we do get a close equivalence between the two ratios.

p. 119: Footnote: Current value of mean Earth-moon distance—Kopal, p. 71.

p. 119: Twenty-five weary lightyears to Vega—Ridpath, p. 108 (section on the constellation Lyra).

p. 120: Vega's parallax is about 0.13 arc-second—Determined from Norton, p. 112 (Table 49: Distances in parsecs and light years equivalent to any parallax . . .").

p. 120: "Dividing the line so defined on the fixed rod"—Ptolemy, p. 166 (V.12).

p. 120: "While true, this seems an odd statement"—Jensen, n. to ms. p. 133.

p. 120: Stellar parallax remained unobserved until 1838—Hartmann, p. 237.

p. 121: "They are hidden at the time they are in conjunction"—Copernicus, p. 732 (V.1).

p. 121: An outer planet at its closest to Earth is in opposition to the sun—I have changed "in conjunction to the Earth" and "in opposition to it" to the current phraseology, per Jensen (n. to ms. p. 134), but I do so slightly uneasily, since Kopal (p. 146) insists that "the outer of the two planets is then in *opposition* to the inner, while the inner is in *inferior conjunction* with the outer" (italics in original).

p. 121: Planets separated by 180 degrees on the celestial wheel fall into inauspicious contradiction—Seligmann, p. 252.

p. 121: Mercury's conjunction distances from us: 136 and 50 million miles—Davidson, pp. 118, 113.

p. 123: Footnote: No parallax correction required—Davies, pp. 101, 109.

p. 123: As Copernicus implied, "they are hidden,"—All the same, he advises us, and the figure suspiciously resembles Venus's maximum angle of elongation—another example of the many miseries of Copernican terminology—that 45° 57' "is the additiosubtraction of the great parallax of Venus according to calculations." Copernicus, p. 823 (VI.5).

p. 124: "The sun as a natural dividing line between"—Ptolemy, p. 270 (IX.I).

p. 124: "Are not occulted by the approach of the sun"—Copernicus, p. 584 (II.13).

p. 124: "Venus is drawing in toward the sun"—Ball, p. 24.

p. 124: Definition of angular elongation—Angular elongation may also refer to the angle between a planet and its satellite as seen from Earth.

p. 124: The other planets can cycle through any and all angles.—Hartmann, p. 102.

p. 124: "How unconvincing is Ptolemy's argument"—Copernicus, p. 523 (I.10).

p. 126: "This distinction between the two kinds of planets"—Ptolemy, p. 472 (Appendix B).

p. 126: Venusian orbital eccentricity of 0.007—Jacobsen, p. 20.

p. 126: Ptolemy's translator: "the sun should always be on a line with the center of the epicycle"—In my opinion this last clause is not quite correct; as we see in the diagram of Ptolemy's Venusian orbit, the line from the epicycle's center parallels the line from Earth to the mean sun.

p. 126: Source of that same quotation—Copernicus, p. 488.

p. 128: Bode's Rule remains inexplicable—Hartmann, p. 112.

p. 128: "Completely determines the relative sizes of the epicycle and deferent."—Kuhn, p. 65.

p. 128: "Then what will they say is contained in all this space"—

Copernicus, p. 523 (I.10). "Copernicus did not blush to admit . . . that the orb of the Earth, with its satellite, should be placed in the large space that separates the convex part of the sphere of Venus from the concave part of the sphere of Mars" (Dugas, p. 37).

p. 128: "With it being so big"—Jensen, n. to ms. p. 140.

p. 130: "The movement of Venus"—Copernicus, p. 780 (V.22).

p. 130: "Unlike Ptolemy, whose aim was to keep the planes"—Jacobsen, p. 112.

p. 132: "As the followers of Plato suppose"—Copernicus, p. 521 (I.10). One commentator insists: "Through a misunderstanding of this passage it has often been said that Copernicus predicted the discovery of the phases of Venus and Mercury when means of improving human vision were found" (Copernicus [Rosen], vol. 2, p. 355 [note]). But he holds a solitary view, and I fail to see why believing in such a prediction would be a misunderstanding.

p. 132: Logically impossible to see more than half of Venus at any one time—Kopal, p. 43.

p. 133: In 1611, Galileo's telescope will bear him out—In his dialogue of 1632 (quoted in Rosen, p. 89), Galileo further reasons: "I do not see how it is possible to avoid the statement that this planet revolves around the sun . . . This circle cannot embrace the Earth, because [in that case] Venus would sometimes become opposite the sun. Nor can the circle be below the sun, because Venus would appear horned near both its conjunctions with the sun. Nor can the circle be above the sun, because the planet would always look round and never horned."

p. 133: "The ultimate safeguard of all scientific method"—Amsden, p. 39.

p. 133: "But what about Venus's phases"—Rosen, p. 91.

p. 133: "Copernicus expressed no opinion"—Ibid., p. 93.

p. 133: "Another and greater difficulty does Venus exhibit"—Brophy and Paolucci, p. 91 (Galileo, "Dialogue on the Great World Systems," 1632).

p. 134: "But now the Telescope manifestly shows these horns"—Ibid., p. 97 (Galileo, "Dialogue on the Great World Systems," 1632).

p. 134: Modern figures for the variation in Venus's apparent diameter—Kopal, p. 41.

p. 136: "Now we shall fulfill our promise"—Copernicus, p. 557 (II, introductory para.).

p. 136: Kuhn remarks that had *Revolutions* ended with Book I—Kuhn, p. 184.

p. 136: The "detailed technical study" of Books II through VI—Loc. cit.

p. 137: "But these two circles which have their centers"—Copernicus, p. 558 (II.1).

p. 137: (And this again derives from *The Almagest*)—Ptolemy, p. 25 (I.12). This device differs from Copernicus's only in insignificant ways.

p. 137: Each of whose sides is about two meters long—Copernicus, p. 558 (II.2). The translator actually writes "about five or six feet on a side," but the units in my book are generally metric and since Copernicus's main criterion here is that "the square is of sufficient area to admit being divided into sections," my substitution ought to serve.

p. 138: Footnote: Ptolemy measures the quadrant's arc at between 47° 40' and 47° 45'—Op. cit., p. 26 (I.12).

p. 139: "The angle is right where the meridian circle"—Copernicus, p. 560 (II.3).

p. 139: Tables in II.3–4—They allow us to calculate, for instance, the declination—Ibid., p. 563 (II.3).

p. 139: "For any given altitude of the sun"—Ibid., p. 567 (II.6).

p. 139: "The differences between right and oblique ascensions"—Ibid., p. 576 (II.9).

p. 139: "When some degree of the ecliptic"—Ibid., p. 577 (II.9).

p. 139: "Wherefore *EN* is half the chord"—Ibid., p. 570 (II.6).

p. 139: "Hence the risings and settings"—Ibid., p. 570 (II.6).

p. 139: "If we take the right ascension"—Ibid., p. 583 (II.11).

p. 140: "For without the moon there is no way"—Ibid., p. 587 (II.14)

p. 140: "That most outstanding of mathematicians"—Ibid., p. 589 (II.14).

p. 140: "Actually, astronomers still use a system"—Jensen, n. to ms. p. 151.

p. 142: "One should not omit any necessary premise"—Leibniz, p. 27 ("Meditations on Knowledge, Truth, and Ideas," 1684).

p. 143: Maxwell showed experimentally that a lodestone does not draw iron towards it at once—Kaku, pp. 25–27. "The speed at which the 'disturbance' propagates from the magnet to the iron is the speed of light; this was one of Maxwell's hints that light is an electromagnetic wave."—Jensen, n. to ms. p. 153.

p. 143: "A formal statement that is assumed"—Leithold, p. 4.

p. 144: "Thou created Heaven and Earth"—Augustine, p. 101 (*Confessions*, XII.7).

p. 144: "Before the tenth century"—Kuhn, p. 106.

p. 145: "The first cause of the first movement"—Ptolemy, pp. 5–6 (I.1).

p. 145: "We worship truth itself"—Biechler, pp. 85–86.

p. 145: "However the science of cosmology should develop"—Peach, p. 107. Italics mine.

p. 146: "The sun had risen on the Earth when Lot came to Zo'ar."—Genesis 19.23.

p. 146: "We are speaking in the usual manner"—Copernicus, p. 557 (II, introductory remarks). Three centuries before him, Saint Thomas Aquinas himself had not hesitated to grant the same point.

p. 146: "In the case of the horizon"—Copernicus, p. 559 (II.3).

p. 146: The seven stars in Christ's right hand—Revelations 1.16.

p. 147: "One cannot say anything about an initial creation"—Peach, p. 106.

p. 147: "Shall I and your mother"—Genesis 37.9–10.

p. 147: "When the morning stars sang together,"—Job 38.4–7.

p. 148: "In reading his words throughout *Revolutions*"—Jacobsen, p. 108.

p. 149: "Aspect of knowledge called *proof*"—Cardano, pp. 213–24.

p. 149: "However, in this field of understanding"—Loc. cit.

p. 149: "I shall pursue this analogy in my further cosmological work."—Koestler, p. 125. For an example of that pursuit, see Kepler, pp. 853–84 (IV.I.1).

p. 149: "In the center of all rests the sun"—Ibid., p. 527 (I.10).

p. 149: "How exceedingly fine"—Copernicus, p. 529 (I.10).

p. 149: Why not consider *Revolutions* itself a work of Scriptural astronomy?—This game can be played endlessly. For instance, why not posit, convincingly or not, that secular society is to *Revolutions* as *Revolutions* was to Scriptural astronomy? "The system of checks and balances incorporated in the Constitution of the United States, for example, was intended to give the new American society the same sort of stability in the presence of disruptive forces that the exact compensation of inertial forces and gravitational attraction had given to the Newtonian solar system" (Kuhn, p. 263).

p. 150: "I have summed up the history of 2,454 years"—Chemnitz, p. 51.

p. 150: "Our reason exalts itself"—Ibid., p. 47.

p. 150: Footnote: "Our opinion is that the Scriptures"—Brophy and Paolucci, p. 48 ("Notes Indicating Galileo's Views on the Arguments of Cardinal Bellarmine's Letter," 1615).

p. 151: "Not so much in the astronomical doctrine as such"—A. C. B. Lovell, p. 10.

p. 151: "Be fruitful and multiply"—Genesis 1.26–27.

p. 151: He certainly did not give us dominion over Heaven—See, for instance, Isaiah 14.13–15.

p. 151: "Let there be lights in the firmament"—Genesis 1.14ff.

p. 151: Do not sell grain during the new moon—Amos 8.5.

p. 151: Nor by any means are we to worship them—2 Kings 23.5.

p. 151: "Thou hast made the moon to mark the seasons"—Psalms 104.19.

p. 151: The One "who shakes the Earth"—Job 9.6–8.

p. 152: "The sun will be darkened"—Matthew 24.29.

p. 152: "Copernicus would have spoken correctly"—Quoted in Rosen, pp. 158–59.

p. 152: "The sun also rises"—Ecclesiastes 1.5. I have altered "rises" to "also rises," since that slightly older form is more a part of our language (e.g., Hemingway's *The Sun Also Rises*) than the revised standard.

p. 152: "So the sun turned back on the dial"—Isaiah 38.8–9.

p. 152: God "makes His sun to rise"—Matthew 5.45.

p. 153: "Then Joshua spoke to the Lord"—Joshua 10.12–13.

p. 153: "The Earth is encompassed by seven heavens"—Ladner, p. 81.

p. 153: The Pseudo-Dionysius the Aeropagite postulates nine ranks—Ibid., p. 92.

p. 153: "Transcendence over every earthly defect"—Ibid., p. 93.

p. 154: "That special mathematical theory would most readily prepare the way"—Ptolemy, p. 6 (I.1).

p. 154: Saint Augustine uses the fact that oil always rises—Augustine, p. 1012 (*Confessions*, XIII.9).

p. 154: "Not all flesh is alike"—1 Corinthians 15.39–41.

p. 155: "We believe it is the necessary purpose"—Ptolemy, p. 83 (III.2).

p. 155: The Christian God is said to have originated all things—After Ladner, p. 65.

p. 155: "A vast republic of gods"—Santillana, p. 60.

p. 156: "In those things which it does contain it is obscure and ambiguous"—Chemnitz, p. 46.

p. 156: "Indeed we declare, announce, and define"—Ross and McLaughlin, p. 236 (Pope Boniface VIII, from the Bull "Unam Sanctam," excerpted here as "The Superiority of the Spiritual Authority").

p. 156: First capital punishment for heresy—Lea, vol. 1, p. 385.

p. 157: "Prove that what I advanced were errors"—Bonneschose, p. 74.

p. 157: He dares to argue against the Cardinal of Florence—Ibid., p. 122.

p. 157: "That the origin of councils is derived from the Apostolic Synod"—Hefele, p. 1.

p. 157: There are eight kinds of councils plus remainder of para.—Summary from Hefele, pp. 1, 3, 5, 55, 63–64.

p. 158: The Bishop of Chioggia proposes—Waterworth, p. lxxxiiiff.

p. 158: "The license of interpretation"—Ibid., p. xsc.

p. 158: The Synod "venerates with an equal affection"—Ibid., p. 18.

p. 159: "We should believe that the beginning of the world took place specifically at the equinox."—Bede, p. 24.

p. 159: "I myself being once desirous to know"—*De ascencione mentis in Deum*, quoted in Santillana, p. 85.

p. 160: "As you [are] aware," he threatens Galileo's go-between—Quoted in Santillana, p. 99.

p. 161: Tiberius cast horoscopes to see into enemy plots—Seligmann, p. 68.

p. 161: The Mesopotamians who calculated from the position of the Libra the Balance—Ibid., p. 6.

p. 161: "Moreover," pursues Copernicus, "an irregular movement"—Copernicus, p. 622 (III.1).

p. 161: "The head of the constellation of Aries"—Loc. cit.

p. 162: "For the sake of a cause for these facts"—Ibid., pp. 622–23 (III.1).

p. 162: "The equality of the solar year is more correctly measured from the Sphere of the Fixed Stars"—Ibid., p. 648 (III.13).

p. 162: Copernicus adds 28 seconds—Ibid., p. 649 (loc. cit.).

p. 162: Our computation of 9.5 seconds—Jensen, n. to ms. p. 181. Jensen should also be credited for the immediately following sentiment "but still impressively accurate."

p. 163: "He knew how to explain the hidden causes"—Bienkowska, p. 19 (quoting Wawrzyniec Korwin).

p. 163: "Venereal and mercurial"—Barret, p. 151.

p. 163: And which lies on the celestial equator—Ridpath, p. 140.

p. 163: "Wherefore in the year of Our Lord 1525"—Copernicus, p. 624 (III.2).

p. 164: The equinoxes and solstices moved eastward about one degree per century in Ptolemy's time—Ptolemy notes (p. 233 [VII.4]) that the Sphere of the Fixed Stars turns one degree eastward every century and that the axis of this rotation runs between the ecliptic poles.

p. 164: And one degree per seventy-one years thereafter—Aaboe (p. 97) mentions Ptolemy's "Arabic and other successors," who further refined the precession of the equinoxes from one degree per century to one degree per 71.6 years.

p. 164: "For in the 266 years"—Copernicus, p. 625 (III.2).

p. 164: "Two reciprocal movements belonging wholly to the poles"—Ibid., pp. 626–27 (III.3).

p. 165: "We have to show that when the twin movements of circles *GHD* and *CFE*"—Ibid., p. 628 (III.4).

p. 165: "If the two circles"—Original manuscript version of III. 4+. Koestler, p. 217.

p. 166: "The odd fact is that Copernicus had hit on the ellipse"—Ibid.

p. 166: He finds that its complete cycle lasts 3434 years—Copernicus, pp. 632–33 (III.6).

p. 166: Jacobsen will point out that obliquity is not accurate as *Revolutions* defines it—"Thus obliquation is seen to have the following definition: the correction to the geocentric latitude resulting from a sinusoidally varying oscillation (period equal to the sidereal period of the planet, amplitude itself sinusoidal with the synodic period) of the plane of the planet orbit, about a mean inclination of the orbital plane and referred to an erroneous line of nodes passing through the mean position of the Sun (not through the Sun itself!)"—Jacobsen, p. 142.

p. 166: "Let our problem be to find the true position of the spring equinox"—Copernicus, p. 645 (III.12).

p. 166: "Can be understood two ways, either through an eccentric circle"—Ibid., p. 653 (III.15).

p. 167: For the Earth, as for the sun, moon and five known planets, "all their apparent irregularities"—Ptolemy, p. 270 (IX.2).

p. 167: Gas giants such as Jupiter and Saturn rarely surpass orbital eccentricities of 0.1—Kopal, p. 116.

p. 167: Other orbital eccentricities in same para.—*The Times Atlas*, p. 26; planet table.

p. 167: "Additions-or-subtractions arising from eccentric orbital circle"—Copernicus, p. 670 (III.24).

p. 167: "And so the explanation of the appearance of the sun"—Ibid., p. 671 (III.25). Italics mine.

p. 168: "As a person," concludes Arthur Koestler—Koestler, p. 677.

p. 168: "His character remains silent for us to the end"—Gingerich, ed., p. 171.

p. 168: "He eschewed all ordinary society"—Ball, p. 32.

p. 169: "The astronomy of my lord and teacher"—Copernicus Anniversary National Organizing Committee, p. 10.

p. 169: In a plan from the following century, Torun resembles a broken star—Bietkowski and Zonn, p. 13.

p. 169: The Vistula's frequent floods required the relocation of the place—Ross and McLaughlin, pp. 426–27 (*Ordensritter und Kirchenfürsten*, thirteenth century).

p. 169: "When in doubt, he preferred to remain silent"—Gingerich, ed., p. 172 (Bronowski).

p. 170: No evidence that he ever became an ordained priest—Bienkowska, p. 17.

p. 170: His father died in the same year when the Alphonsine Tables were printed—Rosen, p. 33.

p. 170: "In medicine he was renowned as the second Aesculapius"—Quoted by the Copernicus Anniversary National Organizing Committee, p. 33

p. 170: So a contemporary plan depicts this two-membered constellation (Cracow)—Bietkowski and Zonn, p. 35.

p. 171: He orders blank pages bound in with his copy, plus reason for same—Rosen, pp. 33, 40.

p. 171: List of readings and lectures—Bienkowska, p. 16.

p. 171: Following the approved course of a young Polish gentleman—Milosz, p. 27, who adds: "Such young gentlemen usually attended northern Italian universities."

p. 171: The lunar occultation of Aldebaran in 1497—The dates of various eclipses, observations, etc., are as I have taken them from the pages of *Revolutions*.

p. 171: The moon's parallax refuses to correspond to Ptolemy's predictions—Bienkowska, p. 17.

p. 171: He first begins to mull over Pythagoras's heliocentric notion—Koestler, p. 225.

p. 171: Neoplatonists preach that a finite Aristotelian universe would limit God's perfection—Kuhn, p. 132.

p. 172: "In the center of all rests the sun"—Copernicus, pp. 526–27 (I.10).

p. 172: "These groupings lean on a broken reed"—Rosen, p. 157.

p. 172: "We made careful observations of the other eclipse"—Copernicus, p. 703 (IV.14).

p. 172: "Led a solitary life"—Milosz, p. 38.

p. 173: "There is no one center"—Copernicus (Rosen), vol. 3, p. 81 (*Commentariolus*).

p. 173: Poland gets nicknamed *Paradisus Hereticorum*—Milosz, p. 29.

p. 173: Gynopolis as he playfully Greekifies it—Copernicus (Rosen), vol. 3, p. 333 (letter of 1538).

p. 173: "We ourselves made observations of a second position"—Copernicus, p. 782 (V.24).

p. 174: R. F. Matlak gives him credit—Copernicus Anniversary National Organizing Committee, p. 28.

p. 174: He defends against epidemics and secures the water supply for Warmia—Ibid., p. 35.

p. 174: Alexander Rytel, M.D., believes him directly responsible—Ibid., p. 37.

p. 174: "Venerable and worshipful gentlemen"—Copernicus (Rosen), vol. 3, p. 309 (letter of 22 October 1518).

p. 174: Jacobsen believes him to have been "perfectly happy"—Jacobsen, p. 109.

p. 175: Andrew's leprosy now far advanced—Copernicus (Rosen), vol. 3, p. 321.

p. 175: "The scorn which I had reason to fear"—Quoted without clear attribution in Bienkowska, p. 25. Probably the letter of dedication.

p. 175: "I see therefore that . . . my reward for affection is to be hated"—Copernicus (Rosen), vol. 3, p. 312.

p. 175: "Being a great astronomer, (Werner) is not aware"—Copernicus (Rosen), vol. 3, p. 148.

p. 175: "For what else shall we be doing except giving a hold"—Copernicus, p. 678 (IV.2).

p. 175: Having given advice on calendar reform—Bienkowska, p. 24.

p. 176: Calendar reform impelled him to work out his heliocentric theory—Copernicus Anniversary National Organizing Committee, p. 7.

p. 176: "As the year belongs to the sun"—Copernicus, p. 675 (IV.1).

p. 176: "Copernicus was a dedicated specialist"—Kuhn, p. 184.

p. 176: The cause is angel-power—Rosen, p. 180.

p. 177: "Life's exigencies had prepared Rheticus"—Gingerich, *The Book Nobody Read*, pp. 13, 11.

p. 177: "Faultfinding is of little use"—Copernicus (Rosen), vol. 3, p. 145 (letter against Werner).

p. 178: "This refusal required courage"—Ibid., pp. 323–24.

p. 178: "But since I realize the bad opinion of me"—Ibid., p. 320 (letter of 27 July 1531).

p. 178: "I acknowledge Your Reverend Lordship's"—Ibid., p. 332.

p. 178: "In the case of the other planets I shall try"—Copernicus, p. 511 (Book I, introductory remarks).

p. 179: "It is one of the dreariest"—Koestler, p. 677.

p. 180: "Since in the preceding book"—Copernicus, p. 675 (IV, introductory para.).

p. 180: "She alone of all the planets"—Loc. cit.

p. 180: "For the moon is changeable"—Ibid., p. 727 (IV.29).

p. 180: "Now it is clear that in the middle space of time"—Ibid., p. 685 (IV.5).

p. 181: Copernicus proudly reports that his own numbers—Ibid., p. 693 (IV.6).

p. 181: Copernicus will blame part of the moon's irregularity on its obliquity—Ibid., p. 698 (IV.10).

p. 181: Current obliquity value = 5° 8' 43"—Kopal, p. 70.

p. 181: Luna "bisects the ecliptic"—Ibid., p. 675 (IV.1).

p. 181: "Revolves obliquely around the center"—Copernicus, loc. cit.

p. 181: Every 18 years 223 days, a specific lunar phase will repeat—Nowadays we'd say that the sidereal period of the moon—the time needed to return to its starting point—is about 27 days 7 hours 43 minutes; its synodic period—the interval required for a specific phase to repeat—is slightly more than 2 days longer; and its anomalist month—the time needed to return to the same relative place in its orbit around the Earth—is about an hour longer still. Shortest of all is the draconic month, the time needed for the moon to intersect the ecliptic twice; this is 2 hours less than the sidereal period (Kopal, p. 71). So where does Copernicus's 19-year period come from? By multiplication, 223 synodic months nearly equal 242 draconic ones; either interval works out to 18 years 223 days (Hartmann, p. 24. Kopal gives the figure of 18 years 11 days, which I have buried here since it is farther from Copernicus's value). This so-called "lunar regression cycle" and knowledge of its periodicity go back to Mesopotamian times.

p. 181: The length of time over which the pattern of lunar eclipses repeats—This last sentence was stolen almost verbatim from Jensen, n. to ms. p. 208.

p. 182: "Accordingly the ancients understood"—Ibid., pp. 675–76.

p. 182: "But if this is so, what shall we reply to the axiom"—Ibid., p. 677 (IV.2).

p. 182: "Some other different point"—Loc. cit.

p. 182: "Makes one revolution with respect to the mean position"—Ibid., p. 679 (IV.3).

p. 182: All determined through geometric "additiosubtractions."—Ibid., p. 695 (IV.8).

p. 182: "I say that the lunar appearances agree"—Ibid., p. 679 (IV.3).

p. 182: "The orbit of the moon around the Earth is approximately an ellipse."—Kopal, p. 70.

p. 182: The Earth-moon distance can vary by as much as 14 percent—Ibid., p. 71. I have calculated the percent variation and difference in Earth-moon distance from Kopal's figures.

p. 183: A miserable series of geometric and arithmetical operations—Ibid., p. 722 (IV.25).

p. 183: "We shall have the number of twelfths of the eclipse"—Ibid., p. 728 (IV.31).

p. 183: A formula for determining the duration of any future eclipse—Ibid., p. 729 (IV.32). Eclipses can take place only when conjunctions and oppositions of the moon with the sun occur, which in turn can happen only in the two places where the moon crosses the ecliptic; these points are called nodes. ("Can happen" is not "will happen"; eclipses do not infallibly take place twice a year, because the two nodes must be in line with the sun; and they rotate with a period of 347 days [Hartmann, pp. 22–25, who makes it much simpler].)

p. 183: "From this it will now be apparent how great the distance of the moon from the Earth is"—Ibid., p. 707 (IV.17).

p. 183: 56 times the Earth's radius plus 42 minutes—Ibid., p. 708.

p. 183: "Within a few percent" of the value obtained by Hipparchus—Kopal, pp. 71, 70.

p. 183: Ptolemy's mean value of 59—Op. cit., p. 171 (V.13).

p. 183: "The comparison of the difference in extent of the eclipses"—Copernicus, p. 710 (IV.18).

p. 184: (Apparent mean) diameter of 31' 52"—Kopal, p. 72. I have made the conversion from his 1865 seconds.

p. 184: "Improving" the angular diameter of the moon—Jacobsen, p. 133.

p. 184: Asymmetry keeps one side of the moon eternally facing Earth—Hartmann, p. 79.

p. 184: 1179 times Earth's radius—Copernicus, p. 712 (IV.19).

p. 184: Actual value is 23,455—Since the diameter of Earth is 12,756 kilometers (Hartmann, p. 103, the radius is 6378 kilometers (by the way, Greeley and Batson mislabel this number as the diameter; p. 325; thankfully, Jensen caught this contamination in my original calculations). Earth's orbital eccentricity is low; therefore, the mean Earth-sun distance will serve as a cross-check on Copernicus's distance at apogee (which we would now call aphelion). This distance, 1 astronomical unit, is 149,597,900 kilometers (Kopal, p. 24). The terrestrial radius value goes into 1 AU 23,454.8 times.

p. 184: "And hence the sun will be 6,999 and 62/63 greater"—Copernicus, p. 713 (IV.20).

p. 184: The solar volume, however, he grossly underestimates—*The Times Atlas* (p. 29) gives a lunar volume = 0.02 times that of the Earth. In this case, the ratio of terrestrial to lunar volumes would be 50 to 1. The same source gives the ratio of solar to terrestrial volumes of 1,306,000 (p. 26), making the comparable solar-lunar ratio 653,000,000.

p. 184: For the ratio of solar to lunar volumes he is off by a factor of more than 9000—"This is a direct result of the distance underestimate" between the Earth and the sun, points out Jensen (n. to ms. p. 212). "If he is off by a factor of 20 on the distance, he will underestimate the Sun's radius by the same factor—and thus the volume by that factor cubed: $20 \times 20 \times 20 = 8000$."

p. 185: "Choose not to deny experience"—Dante, *Inferno*, p. 327 (Canto XXVI).

p. 185: "Here, if anywhere, Dante's imagination"—Ibid., p. 331.

p. 185: Dante matches up his universe's three movements with the three angelic Thrones—Dante, *The Banquet*, quoted in Kuhn, p. 114.

p. 186: "Was not eager to publish"—Milosz, p. 38.

p. 186: "Since the newness of the hypotheses"—Copernicus, p. 505, slightly retranslated.

p. 187: Rewarded him with a Greek manuscript—Copernicus Anniversary National Organizing Committee, p. 8.

p. 187: "By the second decade of the seventeenth century"—Kuhn, p. 198.

p. 187: "As for the Catholic hierarchies"—Santillana, p. 3.

p. 188: Pope Paul II never did approve *Revolutions*—Rosen, p. 152.

p. 188: "Nobody accepts it now except Copernicus"—Quoted in Rosen, p. 155.

p. 189: "Has no right to complain about the men with whom he disputed"—Ibid., 157–58.

p. 189: "The Master of the Sacred and Apostolic Palace had planned to condemn his book"—Loc. cit.

p. 190: *Canny* to have died when he published—Rosen (pp. 158–59) writes: "Whereas Galileo was sentenced to prison, Copernicus went scot-free by delaying the publication of his *Revolutions* until he was safely beyond the reach of the ecclesiastical authorities . . . Thereafter the Vatican was so deeply engrossed in the portentous Council of Trent that Copernicus's *Revolutions* escaped further official notice."

p. 190: "The globe of Earth was the centre"—Wells, p. 12.

p. 190: "We have indicated to the best of our ability"—Copernicus, p. 813 (VI, prefatory para.).

p. 191: "The proper annual movements"—Ibid., p. 734 (V.1).

p. 191: "In order for such a vast space"—Copernicus, p. 522 (I.10).

p. 191: "We do not know that this great space"—Ibid., p. 523.

p. 191: "Since Copernicus man has been rolling"—Nietzsche, p. 8 (aphorism 5).

p. 192: "Now we are turning to the movements of the five wandering stars"—Copernicus, p. 732 (V, prefatory para.).

p. 192: "By reason," as Copernicus explains, "of the parallax"—Ibid., p. 732 (V.1).

p. 193: "At a speed equal to the anomalistic passage"—Ptolemy, p. 391 (XII.1).

p. 193: "The parallax due to the great orbital circle of the Earth."—Ibid., p. 807.

p. 193: Mars travels more slowly than we do, and with obliquity to the ecliptic—Hartmann, p. 107.

p. 193: "I say that *when the planet*"—Copernicus, p. 808.

p. 196: "Almost equivalent to the conversion theorem"—Ptolemy, p. 391 (Editorial Note to XII.1).

p. 196: The center of Ptolemy's eccentric corresponds to Copernicus's mean sun—Ibid., p. 392.

p. 196: "The true position of Saturn"—Ibid., p. 732 (V.I).

p. 196: He takes their angles of greatest western and eastern elongation—Ibid., p. 777 (V.20).

p. 197: "Let us pass over in silence"—Ibid., p. 771 (V.16).

p. 197: "The standard procedure involved"—All the various quotations in this discussion are from Kuiper and Middlehurst, pp. 31–33 (Dirk Brouwer and G. M. Clemence, "Orbits and Masses of Planets and Satellites").

p. 198: His error will be 0.55 percent or less—Copernicus Anniversary National Organizing Committee, p. 15.

p. 198: Labored "without much success."—Kopal, p. 2.

p. 198: "Moreover, we took observations"—Copernicus, p. 775 (V.19).

p. 199: Copernicus determines the mean orbital circle of Mars—It will be equal to the ecliptic radius plus 31° 11' (Copernicus, p. 776 [V.19]).

p. 199: "We have compared these three of Ptolemy's observations"—Copernicus, p. 771 (V.16).

p. 199: "Whereby the movement of the planet"—Ibid., p. 770 (V.15).

p. 199: "The known arc *AB*"—Ibid., p. 746 (V.5).

p. 199: Circuit of Martian parallax of 779 days—Ibid., p. 733 (V.1). Characteristically, he then "reduces" this value to "the degrees of a circle"—in this case, the geometric movement of Mars against the Sphere of Fixed Stars: Mars moves 191° 16' 19", 53 sixtieths of seconds, 52 sixtieths of sixtieths each year! (p. 734). "If I could bring my computations to agree with the truth to within ten degrees, I should be as elated as Pythagoras."

p. 199: Current calc. orbital period of Mars = 1.88 years—Greeley and Batson, p. 325 ("Data for Planets and Satellites").

p. 199: Another calculation from the figure . . . plus remainder of this para.—Ptolemy, p. 473 (Appendix B, "The Passage from the Ptolemaic to the Copernican System, and Thence to That of Kepler").

p. 200: "Moreover, we took observations on the conjunction of Mars"—Copernicus, p. 775 (V.19).

p. 200: Drawing an arrow-pierced eccentric circle *ABC*—Ibid., p. 776 (V.19).

p. 200: Mean Martian orbital radius of 1.524 AU—Kopal, p. 55.

p. 200: "So too in the case of Mars"—Copernicus, p. 777 (V.19).

p. 200: Inclination and obliquity figures (the latter are labeled "inclination w.r.t. ecliptic)—*The Times Atlas*, p. 26. They are given in decimal values, without units, but since the Earth's inclination is a familiar

23.45, I have hazarded giving the whole numbers as degrees and rounding the figures to the right of each decimal sign into whole minutes.

p. 200: The equatorial inclination of Mercury and the obliquity of Earth = zero—Loc. cit.

p. 201: "By this composite movement, the planet does not describe a perfect circle"—Copernicus, p. 743 (V.4).

p. 201: "From liability to injury and disparagement"—Ibid., p. 785 (V.25).

p. 201: "When the figure has been drawn"—Loc. cit.

p. 201: To Mercury a period of 88 days—Ibid., p. 786 (V.25).

p. 201: Our own value of 87.969 days—Kopal, p. 67.

p. 201: "Therefore by means of the tables drawn up in this way by us"—Copernicus, p. 806 (V.34).

p. 201: The eccentric circles of those planets become their mean longitudinal orbits about the mean sun—Ptolemy, p. 323 (X.6), trans. fn., nearly verbatim.

p. 202: "Ingenious, though not successful, account"—Jacobsen, p. 141.

p. 202: Saturn's astrological image—E.g., Barret, pp. 15–60.

p. 202: The greatest and least distances of Saturn—Copernicus, p. 759 (V.9).

p. 202: "All these things are in perfect regularity"—Ibid., p. 760 (V.10).

p. 202: The planets sweep round and round to the east—Ibid., p. 807 (V.35).

p. 203: "Neither an original"—Koestler, p. 677.

p. 203: "One last attempt to patch up an outdated machinery"—Ibid.

p. 203: "Copernicus tried to design an essentially Aristotelian system"—Kuhn, p. 84.

p. 204: "Very much like *The Almagest*"—Aaboe, p. 108.

p. 204: "The great treatise of Copernicus"—Santillana, pp. 2–3.

p. 204: "He was presented as a haughty, cold, and aloof man"—Milosz, p. 38.

p. 204: "Actually," laughs a Vatican astronomer—Consolmagno, p. 62.

p. 204: Poor Copernicus is not even mentioned—Davies, p. 6.

p. 205: "There are no other smaller circles"—Kepler, p. 853.

p. 205: "The Philosophy of Copernicus"—Loc. cit.

p. 205: "By help of which false supposition"—Quoted in Kuhn, p. 186.

p. 205: "Basically correct,"—A. C. B. Lovell, p. 12.

p. 205: "We shall take for granted, from the outset"—Herschel, p. 4.

p. 206: "It remains for us to occupy ourselves"—Copernicus, p. 813 (VI, prefatory para.).

p. 207: "Once the Copernican system is supposed"—Ptolemy, p. 472 (Appendix B). Italics mine.

p. 207: The planets' "true positions are said to be known"—Copernicus, p. 813 (VI, prefatory para.).

p. 208: "Accordingly by means of the assumption of the mobility of the Earth"—Loc. cit.

p. 208: Corrected parallax must be added to corrected anomaly if it is "greater than a semicircle," and subtracted otherwise—Copernicus, p. 806 (V.34).

p. 208: "The sought position of the planet"—Loc. cit.

p. 208: "Every digression in latitude is measured from the nodes"—Ibid., p. 814 (VI.1).

p. 209: "So that it almost touched the ecliptic."—Ibid., p. 818 (VI.3).

p. 209: This planet shows the greatest digression of all—Ibid., p. 819 (loc. cit.).

p. 209: Definition of orbital inclination—Simplified from the definition in Norton, p. 188.

p. 209: Definition of apsides—Slightly simplified from the same source, p. 185.

p. 209: Definition of obliquation—After Jacobsen, p. 112 (slightly reworded).

p. 209: Mention of deviation in Ptolemy—Copernicus, p. 814 (VI.1).

p. 209: Definition of deviation—After Jacobsen, p. 112 (slightly reworded).

p. 209: "In the case of things which are undergoing a libration"—Copernicus, p. 815 (VI.2).

p. 210: "Mercury also differs from Venus"—Ibid., p. 818 (VI.2).

p. 210: Inclinations of the planets in the tables—Saturn, Jupiter, Mars: Copernicus, pp. 819–20 (VI.3). Venus and Mercury: ibid., pp. 828–29 (VI.7).

p. 210: Same table, modern values—*The Times Atlas*, p. 26. No units given.

As per my previous procedure for obliquity and inclination, I have converted from decimal values to degrees and round minutes.

p. 210: "The deviations of Venus always remain northern"—Copernicus, p. 837 (VI.9).

p. 211: When Copernicus says "latitude," he might be referring to the latitude of declination, of obliquation or of deviation—Ibid., p. 837 (VI.9).

p. 211: "And the remainder will be the predominant latitude sought for"—Ibid., p. 838 (VI.9).

p. 211: ". . . Except that in the case of Mercury"—Ibid., p. 837 (VI.9).

p. 212: "As far as the appearances"—Ptolemy, p. 12 (I.7).

p. 212: "Astronomy has two ends"—Kepler, p. 852 (IV, "First Book on the Doctrine of the Schemata").

p. 213: Agreeing "fairly well with the angular motions"—Jacobsen, p. 122.

p. 213: "For a time," says Kuhn—Kuhn, p. 229.

p. 213: Reinhold's employs Copernican geometries—Copernicus Anniversary National Organizing Committee, p. 17; Kuhn, pp. 187–88, who adds: "Every man who used the *Prutenic Tables* was at least acquiescing in an implicit Copernicanism."

p. 213: "In every work there are to be observed the situation, motion, and aspect"—Barret, p. 6.

p. 214: The sun has its greatest power in the nineteenth degree of Aries, least in nineteenth degree of Libra—Seligmann, p. 250.

p. 214: Our occultist just quoted remains defiantly evasive—Barret, p. 152.

p. 214: The "Wittenberg interpretation" of Copernicus—Vermij, p. 29.

p. 214: Gregorian calendar indebted to Copernican mathematics—Kuhn, p. 126.

p. 214: Copernicus's tropical year makes Gregorian calendar accurate to one day in three thousand years—Copernicus Anniversary National Organizing Committee, pp. 14–15.

p. 214: "No considerable improvement followed immediately"—Jacobsen, p. 147.

p. 215: *Revolutions* "proposed an important change indeed"—Barzun, p. 192.

p. 215: "It will be necessary for us to assume irregular and corrected movements"—Copernicus, p. 810 (V.36).

p. 215: "His planetary theory could not reproduce the observed values"—Weigert and Zimmermann, entry on Copernicus.

p. 215: Even the astrologers embrace his system—Copernicus Anniversary National Organizing Committee, p. 10.

p. 215: Blaeu represents the Copernican system—Vermij, p. 69.

p. 217: "One of astronomy's greatest triumphs"—Kuhn, p. 261.

p. 217: Herschel discovered Uranus by accident—Kuiper and Middlehurst, p. 12.

p. 218: The translator remarks: "These three points"—Ptolemy, pp. 201–202 (fn. to IX.5).

p. 218: "The great difficulties which beset the monstrous conception"—Ball, p. 39.

p. 218: "Because we can measure an object's position in two dimensions"—Jensen, n. to ms. p. 28.

p. 219: "A kind of skin of the world"—Kepler, p. 855 (IV.I.1).

p. 219: He finally rejects the notion that the spheres . . . to intersect the Sphere of the Sun.—Ibid., pp. 856–57.

p. 219: "The Earth may be a Hawaii in a universe of Siberias"—Hartmann, p. 457.

p. 219: "Rather a geometrical conception"—Herschel, p. 55.

p. 219: $\vec{J}_a = \vec{p}\vec{D}$—Tilley and Thumm, pp. 128–29. Their version gives no vector arrow over the variable D. "I don't know how far the author wants to go in explaining physics to his readers," writes Jensen (n. to ms. p. 257), "but the equation as it is stated is incorrect. The linear momentum p of a planet in orbit is continually changing, since the direction of the planet's motion is changing. Thus, according to this equation, the angular momentum is continually changing as well, at least in direction. But angular momentum is in fact conserved, as the author stated in the previous sentence. The correct equation would be $J = p \times D$, with vector arrows above all three variables. The '×' here denotes a mathematical operation called a cross-product that is a vector multiplication. The cross product $a \times b$ is a vector of ab sin theta (there theta is the angle between the two vectors) and a direction perpendicular to both vectors. In the particular case of an orbit, p and D are always perpendicular to each other, so sin theta is 1 and the magnitude of J is just pD." I have followed Jensen in placing vector arrows over all three quantities, although Jensen is "inclined to say that one could just drop the vector part of this, and say (correctly) that the angular momentum has constant magnitude (size) of pD, or (perhaps more intuitively mvD, where m is the planet's mass and v its orbital speed."

p. 220: "The holy principle of uniformity of motion"—Jacobsen, p. 110 (letter to Pope Paul III, 1542).

p. 221: "The boundary posts of true speculation"—Kepler, p. 849 (note to Book IV).

p. 221: At "the first hour of the night"—Brophy and Paolucci, pp. 26–30 (from Galileo's *Starry Messenger*, 1610).

p. 222: Fifty-nine more had been discovered, twenty-three in 2003—Jensen, n. to ms. p. 260.

p. 222: The critics of Copernicus who "are mightily disturbed"—Brophy and Paolucci, loc. cit.

p. 222: The local orbits of these novelties substantiate Ptolemy's belief in epicycles—Galileo apparently did consider the Jovian moons to prove just that. Copernicus Anniversary National Organizing Committee, p. 24.

p. 222: The heavens, which "are spherical and move spherically"—Ptolemy, p. 7 (I.2).

p. 222: "Now, we have used the things previously demonstrated"—Ibid., p. 184 (V.19).

p. 223: "Binds together the order and magnitude"—Copernicus, p. 732 (V, prefatory para.).

p. 223: "Let the center of the world"—Copernicus, pp. 653–54 (III.15).

p. 223: "Teach the conformation of the whole universe"—Kepler, p. 853.

p. 223: The sun, stars and planets equal the Father, Son and Holy Ghost!—Ibid., pp. 853–54.

p. 223: "For since the other parts of the world are pure"—Copernicus, p. 679 (IIV.3).

p. 223: God "gives the sun for light"—Jeremiah 31.35.

p. 223: "That fool wants to overthrow"—Quoted in Milosz, p. 39.

p. 223: "Prudent sovereigns should bridle"—Copernicus Anniversary National Organizing Committee, p. 8.

p. 224: "Who will venture to place the authority of Copernicus"—Sir Bernard Lovell, p. 53 (Commentary on Genesis, quoted in White).

p. 224: "Never having heard of him"—Rosen, p. 171.

p. 224: Earth's rotational period used to be twenty-one hours, will be sixty days—Both of these pieces of information come from Kopal, pp. 25–26.

p. 224: "A philosophy that sees mankind"—Hartmann, p. 458.

p. 225: "Most people would say: modern science"—Edwards and Pap, p. 525 (Emil L. Fackenheim, "On the Eclipse of God").

p. 225: "When men thought of themselves as miserable sinners"—Barzun, p. 193.

p. 225: "The ultimate triumph of celestial mechanics"—Millikan, p. 39.

p. 225: "Firmament of authority over us"—Augustine, p. 114 (*Confessions*, XII.15); also p. 124 (XIII.34).

p. 225: "We now know that the sun"—Peach, p. 25.

p. 226: "The sun is the principal body"—Kepler, p. 854.

p. 226: "The first bewildered victims"—Santillana, p. 2.

p. 226: Pieter de Bert places an unmoving Earth—Vermij, pp. 26–27.

p. 226: Only Cracow, Oxford and Salamanca never stood against Copernicanism—Copernicus Anniversary National Organizing Committee, p. 16.

p. 228: By one ignorant ass for the sake of others—Quoted in Macpherson, p. 4.

p. 228: "Bruno's attitude was as radical as possible"—Dugas, p. 42.

p. 228: "Bruno was resolutely Copernican"—Ibid., p. 41.

p. 228: "The suppressed rage of the Church"—A. C. B. Lovell, p. 13.

p. 229: "Perhaps your fear in passing judgment"—Ibid.

p. 229: "Oh, Nicolas Copernicus, how great would have been thy joy"—Brophy and Paolucci, p. 97 (Galileo, "Dialogue on the Great World Systems," 1632).

p. 229: "Copernicus finds his main argument"—Vermij, p. 34.

p. 230: They suppress the first chapter of Kepler's first book, plus cause—Lear, pp. 5–7.

p. 230: Kepler delays publication—Until 1609. Ibid., p. 8.

p. 230: "Arouse much unrest"—Ibid., p. 11.

p. 230: "Traditional and conservative elements"—Biechler, p. 3.

p. 230: Once upon a time, "the entire cosmos"—Ladner, p. 66.

p. 231: Kepler's *Epitome* entered onto the Index—Lear, p. 11.

p. 231: "My books are all Copernican"—Ibid., p. 8.

p. 231: Cardinal Schönbeg may have offered to subsidize Copernicus's writings—So we're often told; but see Rosen's caveat above, p. 189, in the "Eighth circle" section of "The Pillars of Hercules."

p. 231: "As to this matter of the Earth turning"—Santillana, p. 25.

p. 232: Copernicus "was comfortable"—Jacobsen, p. 105.

p. 232: "There should be no illusions about the temper"—A. C. B. Lovell, p. 12.

p. 232: "Procured for himself immortal fame"—Santillana, p. 11 (Galileo to Kepler, 1597).

p. 233: "Replete with the pertinacity of the asp"—Ibid., p. 9.

p. 233: "It seems to me that your Reverence"—Quoted in ibid., p. 99.

p. 233: "a system which might well seem the invention of demons"—Lea, vol. 1, p. 561.

p. 233: Statistics of Tedeschi—Op. cit., pp. 100, 105–108.

p. 234: "I have been pronounced by the Holy Office"—Brophy and Paolucci, p. 109 (Galileo's recantation, 1633).

p. 234: "So here we are at last"—Santillana, p. 124.

p. 239: Glossary of Terms: Definitions of albedo, aphelion, apogee, declination, direct rotation, ecliptic, inferior conjunction, magnitude, meridian, node, occultation, opposition, orbital eccentricity, orbital period, perigee, quadrature, secular, superior conjunction, synodic—After Kopal, pp. 144–47, reworded unless enclosed in quotation marks. Definitions of deviation, obliquation, trepidation—After Jacobsen, p. 112 (slightly reworded). Definitions of inclination, latitude, longitude and certain other words—Norton, sometimes simplified. Other definitions come from Hartmann, Taylor, and Greeley and Batson.

p. 241: Ecliptic—Copernicus, p. 557 (II.1).

p. 241: Equinoxes, last sentence of def.—Somewhat after Aaboe, pp. 5–6.

Bibliography

A. Astronomy, mathematics, physical science and history of science
(incl. Copernicus's works)

As can be seen below, my separate citations for Copernicus, Kepler and Ptolemy are mostly from a single volume. I hope that the reader will consider this a convenience and not a bit of puffery.

Dr. Eric Jensen, the dedicatee, prepared some "Astronomy-related comments on William Vollmann's 'Uncentering the Earth.'" I have cited these in the form "Jensen, n. to ms. p. *x*." These documents will be placed with the rest of my archive at Ohio State University.

Asger Aaboe, *Episodes from the Early History of Astronomy* (New York: Springer-Verlag, 2001).

Al-Biruni, *Kitab Tahdid al-Amakin: The Determination of the Coordinates of Cities*, trans. Jamil Ali (Beirut: Centennial Publications, American University of Beirut, 1967; orig. Arabic ed. completed A.D. 1025).

Aristotle, *Works*, vol. 1; in Robert Maynard Hutchins, ed.-in-chief, Great Books of the Western World, vol. 8 (Chicago: Encyclopaedia Britannica, Inc. and the University of Chicago, 1971 repr. of 1952 ed.). The two works within which I have cited are the *Physics*, trans. R. P. Hardie and R. K. Gaye; and *On the Heavens*, trans. J. L. Stocks.

Isaac Asimov, *The Kingdom of the Sun*, rev. ed. (New York: Abelard Schuman, 1960).

Frank Ayres, *Theory and Problems of Plane and Spherical Trigonometry* (New York: Schaum Publishing Co., Schaum's Outline Series, 1954).

Sir Robert Ball, *Great Astronomers* (London: Isbister and Company Ltd., 1895).

Francis Barret, *The Magus, A Complete System of Occult Philosophy* (Seacaucus, New Jersey: The Citadel Press, 1975 pbk repr. of 1967 facsimile ed.; orig. British ed. 1801).

Barbara Bienkowska, ed., *The Scientific World of Copernicus: On the Occasion of the 500th Anniversary of His Birth 1473–1973* (Boston: D. Reidel Publishing Co., Dordrecht-Holland, 1973).

Louis Brand, E.E., Ph.D., Professor of Mathematics, University of Cincinnati, *Vectoral Mechanics* (New York: John Wiley & Sons, Inc., 1930).

James Brophy and Henry Paolucci, *The Achievement of Galileo* (New Haven, Connecticut: College and University Press, 1962).

Nicholas [Nicolaus] Copernicus, *On the Revolutions of the Heavenly Spheres*; in Robert Maynard Hutchins, ed.-in-chief, Great Books of the Western World, vol. 16: *Ptolemy, Copernicus, Kepler* (Chicago: Encyclopaedia Britannica, Inc. and the University of Chicago, 1971 repr. of 1952 ed.). This version of *Revolutions* is trans. by Charles Glenn Wallis. Cited: "Copernicus."

Nicholas Copernicus, complete works, vol. II: *On the Revolutions*, trans. and commentary Edward Rosen, with Erna Hilfstein (Baltimore: Johns Hopkins University Press, 1992 corr. repr. of 1978 ed.). Cited: "Copernicus (Rosen), vol. 2."

Copernicus, complete works, vol. III: *Minor Works*, trans. and commentary Edward Rosen, with Erna Hilfstein (Baltimore: Johns Hopkins University Press, 1992 corr. repr. of 1985 ed.). Cited: "Copernicus (Rosen), vol. 3."

[Copernicus Anniversary National Organizing Committee.] The National Organizing Committee in Australia, *Nicolaus Copernicus Heritage: On the 500th Anniversary of Copernicus* (Melbourne and Victoria: Polish Technical and Professsional Club, 1973).

Martin Davidson, ed., *Astronomy for Everyman* (London: J. M. Dent & Sons Ltd., 1953).

Arthur E. Davies, *Celestial Navigation: A Practical Guide* (Ramsbury, Marlborough, Wiltshire, U.K.: Helmsman Books, an imprint of The Crowood Press Ltd., 1992).

Stéphane Deligeorges, *Le pendule de Foucault au Panthéon: 1851—1902—1995: Le pendule sous "l'oeil de Dieu"* (Paris: Conservatorie National des Arts et Métiers, Musée National des techniques, Éditions du patrimoine, 2002 repr. of 1995 ed.).

René Dugas, *Mechanics in the Seventeenth Century (From the Scholastic Antecedents to Classical Thought)*, trans. Frieda Jacquot (New York: Central Book Company, Inc., copublished with Éditions du Griffon, Neuchatel, Switzerland, 1958).

Storm Dunlap, *Practical Astronomy* (Buffalo, New York: Firefly Books, 2004 repr. of orig. 1985 ed.).

Leo Elders, S.V.D., Ph.D., *Aristotle's Cosmology: A Commentary on the* De Caelo (Assen, The Netherlands: Van Gorcum & Co., 1965).

Owen Gingerich, ed., *The Nature of Scientific Discovery: A Symposium Commemorating the 500th Anniversary of the Birth of Niclaus Copernicus* (Washington, D.C.: Smithsonian Institution Press, 1975). Cited: "Gingerich, ed."

Owen Gingerich, *The Book Nobody Read: Chasing the Revolutions of Nicolaus Copernicus* (New York: Walker & Co., 2004). Cited: "Gingerich, *The Book Nobody Read.*"

Rupert Gleadow, *The Origin of the Zodiac* (New York: Atheneum, 1969).

Ronald Greeley and Raymond Batson, *The Compact NASA Atlas of the Solar System* (New York: Cambridge University Press, 2001; derived from *The NASA Atlas of the Solar System*, 1997).

William K. Hartmann, *Astronomy: The Cosmic Journey*, 2nd ed. (Belmont, California: Wadsworth Publishing Co., 1982 rev. of orig. 1978 ed.).

Sir John F. W. Herschel, Bart., K. H., *Outlines of Astronomy* (New York: D. Appleton & Co., 1872 rev. of orig. 1849 ed.).

Theodor S. Jacobsen, *Planetary Systems from the Ancient Greeks to Kepler* (Seattle: Department of Astronomy, University of Washington, in association with The University of Washington Press, 1999).

Michio Kaku, *Einstein's Cosmos: How Albert Einstein's Vision Transformed Our Understanding of Space and Time* (New York: W. W. Norton & Co., Atlas Books, Great Discoveries ser., 2004).

Johannes Kepler, *Epitome of Copernican Astronomy* and *the Harmonies of the World*; in Robert Maynard Hutchins, ed.-in-chief, Great Books of the Western World, vol. 16: *Ptolemy, Copernicus, Kepler* (Chicago: Encyclopaedia Britannica, Inc. and the University of Chicago, 1971 repr. of 1952 ed.). This version of the *Epitome* is trans. by Charles Glenn Wallis and contains only Books IV and V. This version of the *Harmonies*, by the same trans., contains only Book V. Cited: "Kepler."

Zdeněk Kopal, *The Solar System* (New York: Oxford University Press, 1974 repr. of 1972 ed.).

William J. Kotsch, Rear Admiral, U.S. Navy (retired), *Weather for the Mariner*, 3rd ed. (Annapolis, Maryland: Naval Institute Press, 1983).

Thomas S. Kuhn, *The Copernican Revolution: Planetary Astronomy in the Development of Modern Thought* (Cambridge, Massachusetts: Harvard University Press, 1985 repr. of 1957 ed.).

Gerard P. Kuiper, ed., *The Earth as a Planet* (Chicago: The University of Chicago Press, 1954).

Gerard P. Kuiper, ed., *The Sun* (Chicago: The University of Chicago Press, 1953).

Gerard P. Kuiper and Barbara M. Middlehurst, eds., *Planets and Satellites* (Chicago: The University of Chicago Press, 1961).

John Lear, *Kepler's Dream, with the Full Text and Notes of* Somnium, Sive Astronomia Lunaris, *Johannis Kepleri*, trans. Patricial Frueh Kirkwood (Berkeley: University of California Press, 1965). Cited: "Lear," not "Kepler," since most of what I've drawn on is Lear's long introduction.

Wolfgang Lefèvre, ed., *Picturing Machines 1400–1700* (Cambridge, Massachusetts: The MIT Press, 2004).

G. W. Leibniz, *Philosophical Essays*, trans. Roger Ariew and Daniel Garber (Indianapolis: Hackett Publishing Co., 1989).

Louis Leithold, *The Calculus with Analytic Geometry*, 3rd ed. (New York: Harper & Row, 1976 repr. of 1968 ed.).

A. C. B. Lovell, Prof. of Radio Astronomy in the University of Manchester, *The Individual and the Universe: The BBC Reith Lectures 1958* (New York: Harper & Brothers, 1959).

Sir Bernard Lovell, *Emerging Cosmology* (New York: Columbia University Press, 1981).

Hector Macpherson, *Makers of Astronomy* (London: Oxford at the Clarendon Press, 1933).

Robert Andrews Millikan, *Evolution in Science and Religion* (New Haven: Yale University Press, 1927).

[Arthur P. Norton.] *Norton's Star Atlas and Reference Handbook*, 20th ed., ed. Ian Ridpath (New York: Pi Press, 2004; orig. ed. 1910).

Plato, *Collected Dialogues*, ed. Edith Hamilton and Huntington Cairns (Princeton, New Jersey: Princeton University Press, Bollingen ser. LXXI, 1978 corr. repr. of 1961 ed.; orig. dialogues bef. 348 B.C.). *Republic* trans. Paul Shorey; *Timaeus* trans. Benjamin Jowett.

Ptolemy, *The Almagest*; in Robert Maynard Hutchins, ed.-in-chief, Great Books of the Western World, vol. 16: *Ptolemy, Copernicus, Kepler* (Chicago: Encyclopaedia Britannica, Inc. and the University of Chicago, 1971 repr. of 1952 ed.). This version of *The Almagest* is trans. by R. Catesby Taliaferro. Cited: "Ptolemy."

Ian Ridpath, *Stars and Planets* (New York: Dorling Kindersley, Smithsonian Handbooks, 2002 repr. of 1998 ed.; first British ed. may have been earlier).

Edward Rosen, *Copernicus and His Successors* (London: The Hambledon Press, 1995).

Giorgio de Santillana, *The Crime of Galileo* (Chicago: The University of Chicago Press, 1955).

Kurt Seligmann, *Magic, Supernaturalism and Religion* (New York: Pantheon Books / Random House pbk, 1971; orig. ed. 1948).

Liba Chaia Taub, *Ptolemy's Universe: The Natural Philosophical and Ethical Foundations of Ptolemy's Astronomy* (Chicago: Open Court, 1993).

F. W. Taylor, *The Cambridge Photographic Guide to the Planets* (Cambridge: Cambridge University Press, 2001).

Donald E. Tilley and Walter Thumm, *Physics for College Students with Applications to the Life Sciences* (Menlo Park, California: Cummings Publishing Co., 1974).

The Times Comprehensive Atlas of the World, 10th ed. (London: Times Books, 2001 corr. repr.).

Joshua Trachtenberg, *Jewish Magic and Superstition: A Study in Folk Religion* (New York: Atheneum / A Temple Book, undated repr. of 1939 orig. ed.).

Rienk Vermij, *The Calvinist Copernicans: The Reception of the New Astronomy in the Dutch Republic, 1575–1750* (Amsterdam: Koninklije Nederlandse Akademie van Wetenschappen, 2002).

A. Weigert and H. Zimmermann, *A Concise Encyclopedia of Astronomy*, trans. J. Home Dickinson (New York: American Elsevier Publishing Company, Inc., 1968 trans. of orig. 1967 German ed.).

James A. Weisheipl, OP, Pontifical Institute of Medieval Studies, ed., *Albertus Magnus and the Sciences: Commemorative Essays 1980* (Toronto: Pontifical Institute of Medieval Studies, 1980).

A. M. Welchons and W. R. Krickenberger, *Trigonometry with Tables* (Chicago: Ginn & Co., 1954).

Fred L. Whipple, *Earth, Moon, and Planets*, 3rd ed. (Cambridge, Massachusetts: Harvard University Press, The Harvard Books on Astronomy ser., 1970; orig. ed. 1941).

B. Scriptures and the church

Saint Augustine, *The Confessions, The City of God, On Christian Doctrine*, in Robert Maynard Hutchins, ed.-in-chief, Great Books of the Western World, vol. 18 (Chicago: Encyclopaedia Britannica, Inc. and the University of Chicago, 1971 repr. of 1952 ed.). This version of *The Confessions* is trans. by Edward Bouverie Pusey.

[The Venerable] Bede, *The Reckoning of Time*, trans. and commentary Faith Wallace (Liverpool: Liverpool University Press, Translated Texts for Historians, vol. 29, 1999; orig. ed. *ca.* 725).

[Bible]. *The New Oxford Annotated Bible with the Apocrypha*, rev. standard ed., ed. Herbert G. May and Bruce M. Metzger (New York: Oxford University Press, 1977). Cited simply by chapter and verse.

James E. Biechler, *The Religious Language of Nicholas of Cusa* (Misssoula, Montana: American Academy of Religion and Scholars Press, 1975).

François Paul Émile Boisnormand de Bonneschose, *The Reformers Before the Reformation,* trans. Campbell Mackenzie, B.A. (New York: Harper & Brothers, 1844).

Martin Chemnitz, *Examination of the Council of Trent, Part 1*, trans. Fred Kramer (St. Louis, Missouri: Concordia Publishing House, 1971; orig. Latin ed. 1565–73).

Brother Guy Consolmagno SJ, Vatican Observatory, *Brother Astronomer: Adventures of a Vatican Scientist* (New York: McGraw-Hill, 2000).

The Right Rev. Charles Joseph Hefele, D.D., Bishop of Rottenburg, *A History of the Christian Councils, from the Original Documents, to the Close of the Council of Nicaea, A. D. 325*, trans. and ed. William R. Clark, M. A., 2nd ed., rev. (Edinburgh: T. & T. Clark, 1894; no date given for orig. German ed.; however, it must not have been too much earlier, since trans. was in correspondence with author).

Gerhart B. Ladner, *God, Cosmos and Humankind: The World of Early Christian Symbolism*, trans. Thomas Dunlap (Berkeley: University of California Press, 1995; orig. German ed. 1992).

Henry Charles Lea, *A History of the Inquisition of the Middle Ages in Three Volumes* (New York: Russell & Russell, 1958 repr. of 1887? ed.).

J. V. Peach, *Cosmology and Christianity* (New York: Hawthorn Books, *Twentieth Century Encyclopedia of Catholicism*, vol. 127 under sec. XIII: "Catholicism and Science").

John Tedeschi, *The Prosecution of Heresy: Collected Studies on the Inquisition in Early Modern Italy* (Binghamton, New York: Medieval and Renaissance Texts & Studies, 1991).

The Rev. J. Waterworth, trans. and comp., *The Canons and Decrees of the Sacred and Oecumenical Council of Trent, Celebrated under the Sovereign Pontiffs, Paul III., Julius III., and Pius IV.* (London: C. Dolman, 1848).

C. General history, literature and context

Charles Avery Amsden, *Prehistoric Southwesterners from Basketmaker to Pueblo* (Los Angeles: Southwest Museum, 1976 repr. of orig. 1949 ed.).

Jacques Barzun, *From Dawn to Decadence: 500 Years of Western Cultural Life: 1500 to the Present* (New York: HarperCollins, 2000).

Henryk Bietkowski and Włodzimierz Zonn, *Die Welt des Copernicus* (Warsaw: Verlag Arkady Warszawa; Dresden: Verlag der Kunst, 1973). A book of photographs of the various places where Copernicus lived.

Girolamo Cardano, *The Book of My Life*, trans. Jean Stoner (New York: New York Review Books, 2002; orig. trans. 1929; orig. Latin ed. 1575).

G. G. Coulton, *The Medieval Village* (New York: Dover Publications, 1989 repr. of orig. 1925 ed.; orig. preface del. from repr.).

The Divine Comedy of Dante Alighieri, vol. 1: Inferno, and *vol. 3: Paradiso,* with English trans. and comment John D. Sinclair (New York: Oxford University Press, 1979 repr. of 1939 ed.; orig. Italian text completed shortly before Dante's death in 1321). Cited: Dante, *Inferno,* and Dante, *Paradiso.*

Paul Edwards and Arthur Pap, eds., *A Modern Introduction to Philosophy: Readings from Classical and Contemporary Sources,* 3rd ed. (New York: The Free Press / A Division of Macmillan Publishing Co., Inc., 1973; orig. ed. 1957).

Arthur Koestler, *The Act of Creation* (New York: Dell Books [Laurel]; 1975 repr. of orig. 1964 Macmillan ed.).

Czeslaw Milosz, *The History of Polish Literature,* 2nd ed. (Berkeley: University of California Press, 1983 rev. of 1969 ed.).

Friedrich Nietzsche, *The Will to Power,* trans. Walter Kaufmann and R. J. Hollingdale, ed. Walter Kaufmann (New York: Random House / A Vintage Giant, 1968; orig. German ed. wr. 1883–88).

James Bruce Ross and Mary Martin McLaughlin, eds., *The Portable Medieval Reader* (New York: Penguin, 1977 repr. of 1949 ed.).

Marie-Louise von Franz, *The Cat: A Tale of Feminine Redemption* (Toronto: Inner City Books, Studies in Jungian Psychology by Jungian Analysts ser., 1999).

H. G. Wells, *The Outline of History: The Whole Story of Man,* rev. by Raymond Postgate and G. P. Wells (Garden City, New York: Doubleday & Co., Book Club ed., 1971; orig. ed. 1920), vol. 1.

Acknowledgments

I am very grateful to my editor, Mr. Jesse Cohen, for agreeing to let me learn about Copernicus; to my agent, Ms. Susan Golomb, for making this project happen in the first place; to Susan's assistant, Ms. Kim Goldstein, for reducing my procedural burdens with her cheerful intelligence and industry. I kiss her fingernails. My copy editor, Ms. Mary N. Babcock, politely saved me from myself on many occasions. Mr. Adrian Kitzinger did a beautiful job of making over my diagrams, and Dr. Jensen also made some valuable revisions. Ms. Leslie DeVries was a source of astronomical encouragement. I also wish to thank Mr. John DeCaire for fine companionship and chauffering on Copernicus-related library excursions.